NETWORKED CRIME

Does the Digital Make the Difference?

Matthew David

BRISTOL
UNIVERSITY
PRESS

First published in Great Britain in 2023 by

Bristol University Press
University of Bristol
1–9 Old Park Hill
Bristol
BS2 8BB
UK
t: +44 (0)117 374 6645
e: bup-info@bristol.ac.uk

Details of international sales and distribution partners are available at bristoluniversitypress.co.uk

© Bristol University Press 2023

British Library Cataloguing in Publication Data
A catalogue record for this book is available from the British Library

ISBN 978-1-5292-1810-7 hardcover
ISBN 978-1-5292-1811-4 paperback
ISBN 978-1-5292-1812-1 ePub
ISBN 978-1-5292-1813-8 ePdf

The right of Matthew David to be identified as author of this work has been asserted by him in accordance with the Copyright, Designs and Patents Act 1988.

Cover design: Andrew Corbett
Front cover image: 123rf/ktsdesign
Bristol University Press use environmentally responsible print partners.
Printed in Great Britain by CPI Group (UK) Ltd, Croydon, CR0 4YY

FSC
www.fsc.org
MIX
Paper | Supporting
responsible forestry
FSC® C013604

To the one thousand or more students who have taken my cyber crime module over the years. You have taught me most of what I know about the topic now, and no doubt quite a lot that I've forgotten.

Contents

About the Author

Matthew David is Associate Professor of Sociology at Durham University in the UK. He has published extensively in the fields of science, technology and (in particular) the digital. He is author of *Knowledge Lost in Information* (with David Zeitlyn and Jane Bex, British Library, 1999), *Science in Society* (Palgrave, 2005), *Peer to Peer and the Music Industry: The Criminalization of Sharing* (Theory, Culture and Society Monographs, Sage, 2010), *Owning the World of Ideas* (with Debbie Halbert, Sage, 2015), and *Sharing: Crime against Capitalism* (Polity, 2017); and co-editor of *The Sage Handbook of Intellectual Property* (with Debbie Halbert, Sage, 2014).

Acknowledgements

I would like to thank Naci Akdemir, Elham Amini, Wajd Beshara, Jack Birmingham, Vaughan Bryan, Grace Carroll, Manuel Castells, Lucy Cawson, Stan Cohen, Will Craige, Bill Dutton, Angela Gage, Shubha Ghosh, Debbie Halbert, Anahita Hoose, Emma Jespersen, Jamie Kirkhope, Andrew Kirton, Bethany Klein, Chris Lawless, Christopher May, Fadhila Mazanderani, Mark McCormack, Linda McKie, Cynthia Meersohn, Pete Millward, James Milton, Tiago Moreira, Mike Presdee, Dawn Preston, Amanda Rohloff, Chris Rojek, Jessica Silbey, Laura Skilling, Paul Stevens, Becky Taylor, John Tehranian, Rebecca Tomlinson, Freya Trand, Siva Vaidhyanathan, Fiona Vera-Gray, Dave Wall, Jon Wistow, Keming Yang, Majid Yar, Peter Yu, David Zeitlyn and Chong Zhang.

Preface

I was born in the late 1960s, just before the 1969 launch of the Advanced Research Projects Agency Network (ARPANET), which laid the foundation for today's internet. Today's mobile network youth might find it shocking that, aged ten, my idea of staying connected when cycling to nearby towns and villages was to have a ten pence piece in my pocket to phone home in the event of an emergency. My cycling trips netted me some free car and travel brochures, half a pornographic magazine and a semi-functioning Binatone tennis computer game console. Whether today's young have had their brains addled by some much easier access to such content I have some doubt (evidence is that today's youth are for the most part rather a progressive bunch), but I do recommend a bit more cycling. In the 1980s I marvelled at the arrival of the compact disc, even though I could not afford to buy one at the time. Still, eventually, their unencrypted digital formatting meant others would not have to pay, as the content of these discs would be uploaded and circulated online for nothing. My nephew still thinks CDs are coffee-cup coasters. My time at university in the late 1980s and 1990s witnessed the rollout of online public access catalogues (that then grew into full-text content services), and then digital swipe cards (and usernames/passwords); the latter attempting to regulate access to the content made visible by the former. The millennium bug came and went or may never have come at all. 9/11 happened. Digital networks were said to have helped a new generation of terrorists, but also became ever more central to efforts to prevent terrorism. The early 2000s also saw chip and pin (as well as various new forms of digital encryption) kill off earlier forms of fraud, but in enabling the expansion of e-commerce also incited a new cat-and-mouse battle between encryption and intrusion. As an early de-adopter of social media channels perhaps I am immune to most of the disinformation that is going around, but perhaps that is itself fake news and I have filters and bubbles all my own. It may be possible for the internet to help steal my identity, but perhaps one day it might give it back if I ever find myself forgetting what I never took notes on at the time (but one digital system or another did).

1

Introduction

On 12 August 2021, a man in Keyham, Plymouth, UK, killed his mother and four other people (including a three-year-old child) with a shotgun, before killing himself. His gun had been removed and his licence revoked in 2020 after a report of threatening behaviour. Both had then been returned in July 2021. It is claimed that the killer had been inspired by online InCel (involuntary celibate) forums, where marginalized men rant about women who reject them. Keyham is also one of the poorest neighbourhoods in England and is situated next to a large military base. Indeed, many factors will have contributed to the killer's choices. 'Mass shootings' are rare in the UK (one occurring every decade or so). Countries with higher incidents of 'mass shootings' have similar internet access to other countries not experiencing mass shootings, but do have higher levels of social inequality and poverty, and greater access to guns. Explaining rare events is complicated. Most visitors to InCel forums do not become killers. Most socially excluded people do not turn to crime. Perhaps the internet 'contributed' to this person's actions; however, if the killer's online rants had been identified, his licence might not have been returned. Yet, mass shootings happened before the internet. They certainly have not stopped because of it. However, checking a person's online profile for 'hate speech' might in future make it easier to identify people who should not have access to guns. As such, if digital networks can promote hate, they might also be used to reduce violence. Simply blaming the internet may blind us to possible solutions to problems, solutions to which digital networks might also contribute.

In 2016, a man dressed as a 'killer clown' robbed a post office in Middleton on Teesdale in County Durham, England; this was the first 'case' of 'killer clown' crime in the United Kingdom that year, allegedly following on from an internet publicized spate of 'killer clown attacks' that had begun earlier that year, originating in the United States. Can the internet make you go bad? Perhaps the person concerned chose a 'killer clown' outfit in which to commit the robbery because such disguises were 'trending' at the time, but it is less likely that the perpetrator chose to rob a post office due to online influence.

He's not laughing now: the tears of a killer clown

On the night before Halloween in 2016, shortly after the post office robbery explained earlier, another killer clown incident in the North East of England gathered even more mainstream media attention. This was due to the culprit being caught and put on trial. Eighteen-year-old Michael March dressed as a clown and jumped out in front of a pregnant woman brandishing an axe, which he then scraped along the ground. The woman picked up a brick and threw it at March, who ran off but was later caught by police. The following February, March was sentenced to six months in prison. UK newspapers and television stations ran coverage of the incident, making particular play of the suggestion that March's actions had been caused by the recent spate of killer clown 'pranks' being posted on the internet that had started in the United States earlier in the year. The spate of copy-cat 'pranks' spread to the UK – as did the spate of claims made regarding the viral character of such videos in supposedly triggering imitation and, hence, further criminality.

Where politicians are wont to 'blame the media' for various forms of social disorder and criminality (from violent music lyrics to violent films and video games causing real-world imitation), so mainstream media are wont to blame 'new media' for such things. In the case of killer clowns here discussed, the viral spread of particular images may be said to have some influence on their viewers' actions, but whether this extends only to explain preferences in how to seek attention (such as dressing as a clown), or goes so far as to cause the decision to act with criminal recklessness, is harder to determine.

The history of media is also the history of media-related concerns about the effects of media. Whether the printing press brought about or enabled (afforded) the Protestant Reformation in Western Europe, when print had been the foundation of social order in China before that time, is only the most famous early example of such a discussion. Newspapers blamed cinema and then radio for everything, from mass hysteria (in the case of Orson Welles' 1930s radio adaptation of H.G. Wells' *War of the Worlds*) to the decline of public morals and even the rise of fascism. Later, radio, television and newspapers began blaming the internet for everything, from the rise of child abuse to racism and terrorism. So, allegations of causal influence are part of a longstanding process of passing blame onto the media, or from one type of media to another. The printing press was seen, in Europe at least, to enable new views to be heard, for better or for worse – promoting liberty or licence, depending on your point of view (Man, 2002). The rise of photography, likewise, made the mass production of explicit imagery possible for the first time, just as film later progressed this widening of access to previously restricted content. The video cassette brought such content into private homes, further challenging control and generating fears about 'video nasties'. The rise of digitally distributed media is only the latest example in this history of widening access and supposedly reduced capacity to contain potentially harmful (obscene) content. Whether such an increasingly 'permissive', media-enabled cultural environment generates more crime (or potentially even reduces it) remains as hotly disputed as was the case 150 years ago with the rise of pornographic photographs, or even 500 years ago with the rise of 'free thinkers' who sought to translate and print the bible into the spoken languages of ordinary citizens.

Things: when does technology define the act

'Knife crime' became the highest profile crime in the United Kingdom in 2019, with a significant rise in the number of public killings by means of a bladed weapon. Attention focused upon how far the police were successfully reducing the number of knives being carried in public. However, in the United Kingdom, knives are also by far the most common means of killing when domestic violence becomes murder. When someone is murdered with a knife by an abusive partner or ex-partner, the crime is rarely labelled as 'knife crime'; but when someone is murdered with a knife by someone in the street, the crime is likely to be labelled exactly as 'knife crime'. Sometimes the social circumstances in which a crime occurs define how it is explained (such as in the case of 'domestic violence'), while in other cases the technical artefact by which the crime is committed (such as in the case of 'knife crime') comes to define the act. The same questions arise regarding guns in the United States. Do technical artefacts make a difference? This question is much debated in relation to the internet.

'Guns don't kill people, rappers do'

Whether we focus attention on the means by which crimes are committed or on the motives that lead a person to want to commit a crime is significant, and it also reflects underlying assumptions about the nature of crime itself. Rational choice approaches to the study of crime (Cohen and Felson, 1979) assume the incentive to commit crime exists whenever the reward outweighs the cost of committing a certain kind of action (the effort involved and the likelihood and severity of being caught and punished). From this perspective, the technical ability to commit a criminal act, and the technical ability for law enforcement agents to prevent it are the only significant variables, and, as such, technical affordances become central to accounting for the extent of crime. From the perspective of those that view action as being shaped by cultural forces (Presdee, 2000;

Ferrall, Hayward and Young, 2015), intent to commit an act becomes a variable (as a person is seen as being influenced by socialized motives, not just purely rational cost/benefit). As both factors (technical affordances and socialized incentives) exist, and as their relative significance is hard to determine in any conclusive fashion, the significance of technical tools relative to social causes remains disputed.

The song 'Guns Don't Kill people, Rappers Do' by the musicians Goldie Lookin' Chain parodies this dispute in relation to perhaps the most obvious manifestation of a force-multiplying technology – guns. A force multiplier is any tool whose use increases the scale of effect when combined with other actors and actions. The term actor is often reserved for humans, and non-human factors are usually referred to in relation to them as 'tools'. This distinction is disputed by actor network theory (Latour, 2005), which suggests that humans and non-humans 'act' in their various assemblages such that it is not just humans that use things to fulfil their intentions: objects themselves have effects on how such intentions are formed. In relation to guns, it can be readily argued that, on their own, guns do not determine their use, and, as such, guns cannot be 'blamed' for the actions of their users. However, when statistics are compared between countries, those with high levels of gun ownership tend to have higher levels of murder (such as when comparing the United States and the United Kingdom). Similar age profiles exist for violent crime in both countries, and both countries also have very similar gender profiles for violent crime convictions. Age and gender as characteristics of offender profiles suggest social explanations are significant. However, while these violent crime profiles are similar in their distributions, the scale of death due to criminal violence differs hugely per capita (Roser and Ritchie, 2013). While, in the UK, knives are the single largest means of committing murder, in the US it is guns. In the United States, the per capita murder rate is much higher generally. Can guns, then, be said to make a difference? It would seem so. However, some countries with high gun ownership levels on a parallel with the US (such as Canada and Switzerland) have far lower murder rates than the US. This comparison makes the claim that guns kill people seem less convincing.

Adding the name of a technology to the word 'crime' creates a raft of supposed types of crime: gun crime, knife crime, car crime and cyber crime. In each of these cases, the diversity of crimes lumped together obscures more than it illuminates ('stealing a car' and 'driving while intoxicated' are not intrinsically the same). Moreover, any implication from such terminology that the technology somehow 'causes' criminality should not be assumed in any straightforward fashion. Nevertheless, neither should it be automatically assumed that technology does *not* make a difference (in the sense of more readily affording outcomes, even if it does not determine effects).

4

Does the digital make a difference?

It is easy to identify how digital technologies make certain kinds of **access** to victims easier, and, therefore, that the internet makes such victimization more likely to happen; this may involve dissolving boundaries of distance or overcoming layers of security. On the other hand, it is almost always possible to show that what can be afforded by digital means can and was previously afforded by other means, suggesting that what is now achieved online is only old wine in new bottles. A second line of dispute is between those who argue that, by increasing **concealment** (by means of the 'dark web', blockchain-based communications and currency systems, onion routers [TORs], garlic add-ons and encryption), the internet also increases the likelihood of anti-social behaviour, and those who suggest digital content is more open to surveillance than face-to-face ('real-world') interactions, which might actually deter deviance. A third line of dispute is between those who suggest digital networks increase scope for coordinated regulation, and those who suggest the internet increases the scope for **evasion** beyond geographical, and therefore legal, constraints. A fourth dimension of dispute is between advocates of 'real virtuality', the idea that informational content has real effects in **inciting** action, and those who argue that blaming media content for human behaviour is fundamentally flawed. I refer to these four dimensions of affordance as access, concealment, evasion and incitement.

The problem with 'moral panics'

It is very unlikely that any human invention or development has ever been without some negative consequence, or at least has never created some possibility for use in a negative way. As will be discussed later in this work, the (US Supreme Court) 1984 Sony ruling declared that a video recorder manufacturer could not be held liable for the fact that most users used such a device to record things in a way that breached copyright law. The way something is used cannot automatically be the responsibility of the provider of that technology. Use is not an intrinsic property of the thing itself. The Sony ruling established the principle of 'dual use', such that one possible criminal use should not define an object's legality or otherwise if lawful uses also exist. However, the principle of 'dual use' has been contrasted with the principle of 'predominant use', such that in some instances (like with certain kinds of weapons) certain jurisdictions prohibit specific things on the basis of their most likely or predominant application. If everything has some potential to cause harm, or be used in a harmful way by someone, the question arises, when does reasonable concern about harmful potential use warrant legal prohibition? At what point does reasonable concern become unreasonable 'panic'? The longstanding, but disputed concept of 'moral panic' focuses on this question.

It may well be that some fears are exaggerated, whether from irrational miscalculation or malicious calculation (or a combination of the two). Sometimes, we worry too much, but the possibility of a moral panic about the threat of moral panics should also concern us. Frank Furedi (1997) is certainly correct to warn against too much anxiety and an unbalanced tendency to see the negative potential in everything; but if that is true, it must also be the case that worrying about too much worrying might itself fall into its own trap. A 'culture of fear', as Furedi suggests we are trapped within, is also a culture prone to worrying about having become a culture of fear. It is not a simple matter to work out the appropriate amount of panic. Claiming that a certain level of fear is 'disproportionate' assumes a neutral method of calculating how bad a bad thing is relative to any other kind of harm, but the valuation of acceptable harm is always culturally specific (Beck, 1992; Douglas, 1992).

Within theories of moral panic, there is a tension – between those who worry most about 'panics' caused by dominant groups seeking to increase regulation (Young, 1971; Cohen, 1973; Hall et al, 1978), and those who are most afraid of 'panics' caused within subordinate groups by a lack of regulative authority (Goode and Ben Yahuda, 2009). This tension maps directly onto disputes between those who argue the problem with new media is too much corporate and/or state control, and those who think the problem is that new media is a free-for-all.

Stan Cohen (1985) suggested the possibility of 'good moral panics'. This raises an additional challenge. That we worry about some things today that might be less bad than they were before might suggest 'exaggeration', but if such concerns are part of reducing harms, or reflect higher moral standards today than may have existed in the past, perhaps 'exaggeration' is the wrong word. Still, the question remains, how can we choose between relative risks and benefits when different people attach different values to different (positive and negative) things? This is perhaps particularly salient when considering the case of new media, where the relative balance between technical potential and human choice is especially hard to disentangle.

Moral panics or proportionate concerns

The issue of access

The COVID-19 pandemic saw a very substantial shift in consumer behaviour towards online transactions over the use of 'bricks and mortar' shops. This extended an already existing and ongoing trend in this direction. Both the specific COVID-19 shift and the wider trend have, however, raised concerns regarding the rapid escalation of online fraud. Are digital transactions intrinsically less secure than real-world equivalents, or is the rise of fraud online merely tracking the rise of online transactions in general? Is it a transitional issue or something else? Is 'access' to victims intrinsically greater when potential victims use global networks? Likewise, digital distribution makes access to obscene content instantaneous and global, this, in turn, making access to both the content itself and those participating in such contact easier. Access to such content by children, as well as access to obscene content that involves children, has become a serious cause of concern. As more people conduct a greater part of their lives online, across every dimension of their lives, the scope for all such interactions to be surveilled (whether by state, corporate or other citizens) increases. Are we becoming a surveillance society, where the invasion of privacy becomes impossible to resist? Where once a bully might have waited for you outside the school gates, trolls, stalkers and bullies can now access victims (at least in a virtual sense) via digital devices at all times and wherever the victim goes (at least as long as they remain connected to their digital devices). On all these fronts, has the digital rendered us all permanently accessible to those that would do us harm?

The issue of concealment

The ability to send encrypted communications via digital devices and platforms, it is claimed, has created greater scope for anonymous terrorist communications, the planning and conduct of violent acts, and the funding and promoting of terrorist causes. Interpersonal hate speech can also be sent via anonymous channels. Stalking by tracking someone else's movements via their digital footprint can also enable anonymous intrusion into other people's lives in a way that is hard to identify. Using social media platforms to propagate disinformation (fake news that is deliberately designed to create a false impression) has enabled politically motivated actors to conceal the source of such information and to feed it to targeted audiences in a way that limits the user's knowledge that they are being manipulated. Encrypted channels allow users to circulate criminally obscene content with a much reduced likelihood of being themselves identified or of the content being circulated to be revealed to authorities. Torrent-based, file-sharing services render it very much harder to identify the source of copyright-infringing content, whereas older peer-to-peer services did allow the 'uploader' to be more readily identified. Anonymity allows online scareware extortionists to target victims and to demand payment, such that payment does not create a simple trail back to the identity of the perpetrator. Cryptocurrencies that allow anonymous transactions make online banking almost as invisible as using cash. Once again, the capacity to conceal online has created scope for criminal victimization that limits the capacity to identify the perpetrator, even as digital networks make access to victims easier. Is such asymmetry an intrinsic feature of the online world, or is such a one-sided representation itself an error?

The issue of evasion

In medieval times, reaching Durham Cathedral's front-door knocker allowed a person to claim sanctuary, entitling them to shelter for 40 nights, after which time they would have to either surrender to the authorities outside or pay the Prince Bishop to be rowed downriver and leave England. Such an ability to evade the law by fleeing its jurisdiction can today be supplemented by the ability to continue to access that jurisdiction by digital means, even if that jurisdiction cannot physically reach out and detain the perpetrator. Even when concealment is breached, it is not always possible to successfully prosecute an offender if they remain at a distance; this is despite the fact that such distance does not prevent them from continuing to commit offences by remote access. Moreover, what might be deemed criminally obscene in one jurisdiction might not be illegal in another jurisdiction, and such content can relatively easily be transmitted digitally across physical borders. Livestreaming and file-sharing services hosting copyright-infringing content may set up their servers in jurisdictions that do not hold service providers liable for what users circulate; as such, these service providers can provide services that enable copyright infringement even when such services would be prohibited if located in most countries. The blocking of such services can itself be evaded by means of virtual proxy networks that reroute user access via jurisdictions that do not block access. What might be illegal for one government to do in terms of surveilling its citizens may be lawful for another government to do in relation to foreign citizens, such that cooperation between both states evades the law in each. In a similar vein, free-speech legislation in one country might enable someone to publish content there that would be deemed criminal hate speech in their own country.

The issue of incitement

When someone is attacked or robbed, it is clearly the case that they are the victim, but, 'incitement' involves an initial act that then influences others to commit subsequent criminal acts. Much concern has been expressed over the potential for online content to corrupt those who view it, especially when it is observed that such content may itself be more extreme than content regulated by established media outlets. The example of terrorist radicalization is often taken as the most extreme case in point. If exposure to certain kinds of content can make a person commit acts of politically motivated violence, this might support the view that such content, globally distributed, creates a very severe, new level of threat. Even given the caveat that most people would avoid such content even if they could access it, and that most people who were exposed to such content would not take from it the desire to accord with its message, that some individuals would act on such provocative material does pose a threat. Even if the numbers who would act on such messages is very small, the fact that such content is more readily accessible via online methods does mean the number who are then incited to act is potentially expanded. Most people exposed to violent pornography or child abuse images would be repelled by it, but if even a small proportion are encouraged by it to believe such content is normal and desirable, that is a real problem. Forums promoting particular terrorist groups, misogyny or paedophilia, or that encourage particular forms of self-harm, have fuelled criminal acts, even if such groups do not exist in a vacuum.

The diversity of these four dimensions of dispute can mean that different sides argue past one another, given that they are arguing from different premises and in relation to different examples. Such a 'dialogue of the deaf' is itself compounded by the wide-ranging nature of debates in relation to new-media effects. In this work, this range of debates has been divided into four parts, with each of these, in turn, being split in two. This typology is not a taxonomy of discrete and natural types, as elements of crossover exist, between, for example, fake news and hate crime, between hacking and identity theft, between violent computer games and hate crime, or between pornography and invasion of privacy (such as in the case of non-consensually circulated, so-called 'revenge' pornography). *Hate, obscenity, corruptions of citizenship* and *appropriation* are the four higher order themes around which this work seeks to structure the discussion of digital effects. Against this thematic backdrop, this work will apply David Wall's 'transformation test'. This 'thought experiment' asks whether things would be different if the digital were removed from any particular scenario where currently the digital plays a role in affording action. This work seeks to develop the clarity of Wall's thought experiment by focusing the 'test' upon the four dimensions of possibility as identified: those of access, concealment, evasion and incitement.

David Wall's transformation test: how much is enough?

David Wall (2007, p 34) refers to the 'transformation test' as a 'rule of thumb', or 'a heuristic device', to apply when questioning whether or not the digital really does make a difference. The transformation test simply looks to answer the question; when a particular crime has been committed in a fashion that draws upon the use of digital technologies: 'what is left if those same networked technologies are removed from the equation?' The transformation test is a thought experiment and, as such, seeks to identity the counterfactual situation where all else remains the same but for the existence of networked computers. Might the crime carried out using such technology still be possible, would its possibility remain but be radically diminished, and would the impact of that criminality be reduced in the absence of digital networks?

Transformations may, then, be of three sorts. First there might be 'binary transformations', such that the absence of digital technologies would render certain kinds of crime impossible. This binary transformation only really exists in relation to crimes that relate specifically to computers, or what Wall refers to as 'integrity' crimes, where the crime itself is a 'crime against the machine'. Such 'hacking', where the machine is the target, should be set against crimes where hacking is one step towards accessing other things by means of hacking computers, and this takes us to the second form of potential change, that of 'quantitative transformation', whereby computers enable an escalation of crimes that existed prior to computers. Here, the potential transformation lies in making what could be done by other means easier and potentially possible on a larger scale (in terms of the number of victims, scale of victimization and geographical reach). Finally, there is what might be called 'qualitative transformation', such as when the 'effect' of victimization is made worse over and above simply the quantity of such crimes. One breach of privacy or intellectual property may be more harmful if the content released has a potentially unlimited audience, as might the impact of one act of hate speech or one piece of obscene content.

However, trying to run such a thought experiment is no simple matter. To remove digital networks from the equation might need to be applied symmetrically for the thought experiment to be best conducted. If criminals did not have getaway cars (and later transit vans), perhaps banks would be safer; but when the police did not have fast cars (and transit vans) to chase bank robbers, bank robbers could use horses and carts. Sometimes technologies cancel themselves out. Is digital encryption intrinsically more secure than a physical lock and key, or is it only a matter of relative investment by competing actors? Is digital surveillance intrinsically more intrusive than physical eyes and ears, or, again, is it just a cat-and-mouse question of relative investments? Encryption and surveillance cannot both be getting stronger online, as one would surely negate the other; however, does such a presumed symmetry overlook the reality of limited access to both by most people relative to better-resourced actors intent on exploiting the less well-prepared on both fronts?

The term 'affordance', which highlights potential use rather than the more deterministic concept of 'effect', might be useful here, but the question of whether digital technologies have effects remains important. Just because a technology can be used in different ways and so might cancel out any asymmetrical outcomes that might be attributed to it, this does not mean that such symmetry is always achieved. A tool is a force multiplier – enabling an agent to more easily achieve a goal than would be possible without that tool. Yet, that agent may be met by that same tool opposing them, and the configuration of such tools and their usage creates conditions that 'act' back on those who are said to 'use' them, such that the very distinction between subject (actor) and object (tool) becomes blurred – within a network or assemblage. Transformation in such conditions of emergence is not just more or less of what was the case before. Transformation may become something qualitatively distinct.

Harms and affordances

Within the theme of hate, terrorism and hate crimes stand as incitements to collective violence, while bullying, stalking and trolling promote abuse at a more personalized level. Regarding obscenity, the question of whether content has consequences is most clearly divided between adult and child content. Whether consumption of obscene content causes violent actions is focused primarily on whether adult pornography and violent computer games make consumers more likely to abuse others. Meanwhile, that production of obscene content harms those involved in its production is fundamentally undisputed in relation to child pornography, even as it is contested in relation to content featuring adults. Whether the digital increases or reduces harm relative to pre-digital conditions is, however, disputed in relation to both adult and child pornographic production and consumption. Regarding corruptions of citizenship, dispute hinges first around privacy, and whether citizens are more informed or informed upon today than at any previous time: first, in relation to state and corporate surveillance and counter-surveillance (in the form of whistleblowers and hacktivists); and, second, in relation to fake news, echo chambers and citizen journalism. Finally, concerning appropriation, dispute rages in relation to finance and production – with the risk of financial loss and identity theft being seen as either more or only different today relative to the past, and with online sharing (piracy) seen as a potential threat or boon in connection to music, film, television and software industries.

Separating out different conceptions of what impact the digital might have (accessing, concealing, evading and inciting), and in examining distinct dimensions of digital action (in relation to hate, obscenity, citizenship and appropriation), this work seeks to identify where it makes sense to say the digital makes a difference and why, in other ways, it does not. Where terrorists today may prefer suicide bombing to the long fuse afforded by the internet, they can, nonetheless, recruit, plan, propagandize and fundraise online, even if in doing so they may not be as anonymous or unregulated as they might imagine. Those promoting hate speech, and who bully, stalk and troll online, face similar contradictions. Obscene content (sexual, violent or both) spreads online far further than ever before, crossing borders and challenging law enforcement. Such content may incite violence in particular individuals while not for the vast majority. Global networks challenge regulation, but there have been regulative responses at an international level. Citizens are both better informed and more informed upon than ever before. Ideas of what privacy we should be entitled to have changed. Scope to breach any newly acquired entitlements (or presumed entitlements) has also grown. Fake news is hardly new, and therefore news of its sudden appearance is fake news. Today, traditional media providers are challenged by new information-aggregation services, but in fitting advertiser/state priorities to user self-selection, this challenger does much the same as biased old media did before (Fuchs, 2021). Furthermore, scope to be defrauded, or to have one's forms of identification appropriated, increases as we have more alienable (and, in particular, informational) assets. We have more usernames and passwords by which to be identified. Yet, this is only to say that security online is keeping up with the threat. While sharing online has challenged established business models in the informational economy, sharing is in many respects more efficient, effective and incentivizing than traditional property-based models of protecting informational goods. Profitable adaptations exist in all information sectors.

Affordances to access, conceal, evade and incite are real, but technologies can support multiple forms of each – some of which balance one another, or can be held in balance with sufficient insight, will and attention. Technology should be taken seriously precisely because its consequences can never be taken for granted as we can always do otherwise.

Computer misuses?

Established typologies for understanding cyber crime – in particular, the distinctions between computer-enabled and computer-dependent crimes, and between crimes using, within and against computers – have become unhelpful. Brenner (2007) starts out with a three-way distinction between crimes where computers are the target, the tool or incidental to the crime. This typology follows Grabosky et al's (2001) efforts. However, in both cases, this three-way split is later abandoned, as 'target' crimes (hacking) do not remain sufficiently distinct from 'tool' crimes, and the 'incidental' becomes (well) incidental. With the remaining 'tools' type then becoming all inclusive, more recent works have tended towards a 'cataloguing' approach – with multiple chapters addressing discrete types of crime (see Moore, 2015; Holt et al, 2017; Yar and Steinmetz, 2019). While useful, cataloguing tends to produce very long books with very short overall findings on the global question of 'whether the digital makes a difference' – which is the focus of this book. Lavorgna (2020) offers a five-part typology of crimes against devices, people, deception/coercion, markets and the political, which crosses over to some degree (but not fully) with my own typology of harms. Indeed, it is the contention of this current work that mapping a typology of harms through a distinct typology of affordances offers a more productive structure than just outlining a typology or cataloguing of harms.

The distinction between enabled and dependent cyber crimes was useful in the early days of computing, where 'hacking' could be seen as a distinct form of criminality relative to crimes where computers were used to access people or things. The idea that 'hacking' was a standalone activity has diminished over time, and today pretty much every form of crime involving computers involves some kind of 'hack' (modification of a computer system relative to its creators' intentions). That computer-specific laws (such as the UK Computer Misuse Act 1990) are so rarely used is because, in almost all cases, the more significant criminality related to the use of a computer is covered by other laws (against fraud, extortion, hate crime, terrorism and so on). While there are 'geeks' who seek to 'hack' systems for its own sake, this is a relatively minor issue relative to using computers to achieve other ends. As the only truly 'cyber-dependent' crime is one where the computer is the target, and this only very rarely occurs, the distinction, while valid, is of limited value and may be, in fact, highly unhelpful if it misdirects attention.

Likewise, the distinction between crime against, using and within the computer (Wall, 2007), while useful as an initial heuristic device, soon breaks down in a world where the distinction between the real and the virtual is not straightforward. If crimes using the machine refer to crimes where computers are merely the conduit for illegally accessing other people and things, while crimes within the machine refer to virtual crimes where what is inside the machine is itself illegal (such as obscene materials), this distinction soon breaks down, especially when you consider that money is as virtual as reputation and privacy, and when a piece of content's legality (or otherwise) cannot be separated from its context or use. Is posting a person's private information a 'content' crime if the content is not unlawful, only its release? Online file-sharing breaches intellectual property rights, but, again, it is not the 'content' that is illegal, but rather its distribution. Is using the internet to make and distribute copies a content crime, theft or something else?

Initially, in cyber crime research, there was a strong emphasis on the fact that there were 'real-world' routes to many things that could be done digitally. In principle, however, apart from pure 'hacking' (the act of targeting computers themselves for no other reason than to show it can be done), all crime/harm can be carried out 'by other means' and not just via a computer. The more interaction has moved online, so to a greater degree has access to such interaction meant digital access, and so digital dependence, has grown in practical terms, even as all crimes but hacking are still, in principle, possible by other means. Just as dependence/enabling has blurred, so the distinction between use/within breaks down. While crimes within the machine refer to purely digital content (rather than to the digital means to more tangible 'things'), obscenity, hate, intangible properties and privacy/identity are not just 'inside' the machine, given that they link to tangible things like the body, violence, property rights and personhood. The term hacking, and so the category of 'crimes against the machine' that was supposed to sit between 'crimes using' (computer enabled) and 'crimes within' (computer dependent) the machine, also fragments as a type of crime. Hacking, combining invasion of privacy, trespass, damage, theft, modification and data release, becomes a pervasive means of committing pretty much all crimes online – whether that be fraud, identity theft, cyber warfare, decrypting copyrighted content and circulating it, doxing/trolling, bot-herding, ransomware and so on. In essence, the terms 'computer enabled' and 'computer dependent' (crimes using/within the machine) wither; meanwhile, the term 'crimes against' (hacking) swells to the point of ceasing to be a field, and, rather, it becomes the key element of the 'access' affordance.

David Wall (2007) draws out the development of cyber crime in terms of a transition from, first, 'traditional' crimes becoming subject to digital methods (where, for example, fraud might be carried out by breaking into a standalone computer system) to, second, hybrid forms of cyber crime (where increasingly networks between computers became the target), to, finally, true cyber crime, where elements of computer-related crimes become distributed and automated. Given that full automation is never achieved, and given that no computer is fully standalone anymore, this, again, leaves the hybrid mode of cyber crime as, in a sense, the sum of all cyber crime. Rather than drawing up typologies based on type of use or technical sophistication, this work seeks, then, to approach the question of whether the digital makes a difference from an examination, both of types of harm (hate, obscenity, corruptions of citizenship and appropriation) and of types of affordance (access, concealment, evasion and incitement).

Synopsis

Addressing four distinct digital affordances (access, concealment, evasion and incitement), across four distinct fields of criminal harm (hate, obscenity, corruptions of citizenship and appropriation), this work engages classic debates and cutting-edge research to answer the question: does the digital make a difference in relation to crime; or does old wine in new bottles merely reproduce perennial problems and their solutions, thereby cancelling one another out in the process? In conclusion, the work highlights, firstly that digital affordances are symmetrical in principle, but not always in practice (see next section); and, second, that while symmetrical, digital affordances polarize along such lines of symmetry – creating increased scope for (while not compelling) extreme oppositions. Adopting a comprehensive and integrated approach to the four affordances and four fields of criminal harm, that, taken in isolation, allow commentators to talk past one another, this work will be attractive to anyone looking for entry into a complex and wide-ranging field of enquiry.

Outline

This work engages the question: do digital networks make a difference in the scope, scale and severity of crime, or do they simply reproduce old problems by new means, to no greater effect overall? Debates on this question have been characterized by a dialogue of the deaf. New things happen every day, but it is possible to claim that there is nothing truly new under the sun simply by presenting research that supports that claim. Alternatively, alarming new phenomena can be presented as representing evidence of significantly escalating problems. The solution proposed here is to engage with the full range of what criminological and sociological research there is, in as systematic a fashion as possible. This work sets out a range of criminal harms and a range of affordances by which the digital may or may not intensify or reduce those harms. There are four fields of harm and four affordances. The four fields of harm – each comprised of two subfields – are hate (in the form of so-called cyber terrorism and interpersonal abuse); obscenity (in the form of explicit violent and sexual content depicting adults, and that depicting children); corruptions of citizenship (in relation to privacy and in relation to misinformation); and appropriation (in relation to fraud, virtual robbery and stealing identity, as well as in relation to piracy). There are not just four types of crime, but, rather, these represent a typology of possible types of crime. The four affordances, meanwhile, are those of access, concealment, evasion and incitement (this representing a second typology). This work will undertake an in-depth exploration of the interplay of these affordances within each problem subfield, drawing upon the most up-to-date research. In each case, this work deploys David Wall's 'transformation test', to ask whether things would really be different if digital networks did not exist. The digital has made a difference, but such affordances are never totally asymmetrical, and so *can* be used in different ways, and, as such, *can* cancel themselves out in the process. That is not to say such balancing out *is* always achieved. Will and policy mean that outcomes are a choice, not an outcome determined by the technology. The work draws two conclusions. First, in principle, digital affordances are symmetrical, and can, therefore, cancel one another out; but that does not mean, in practice, that they do. Outcomes are contingent upon action and investment, not compelled by technical necessities. Second, while affordances may be symmetrical, digital networks do extend the potential for amplification, and hence for extremes on both sides. As such, while digital symmetry means there is always scope for balance (and, hence, action, rather than resignation on the basis of some technical inevitability), extension means that imbalances that do arise as a result of inaction can cause greater levels of harm to unfold.

It should be noted that the boundary of the digital is not fixed. Networks come to incorporate more and more things, while more and more things incorporate digital networks. The Internet of Things, rather than just 'the internet' or 'digital communications', means access now means control over physical machines, not just digital 'communications'. The scope of such remote-controlled machines to violate humans extends into all four areas of harm/crime to be discussed in this work.

Empirical range, up–to–date research and theoretical framework

Empirical range – This book is comprehensive in its coverage of the range of social harms that have, or are claimed to have been, intensified by digital network means. As such, topics such as cyber terrorism, fake news, adult pornography and music sharing are addressed precisely because they are disputed forms of harm, legally ambiguous and are disputed as to whether or not digital networks increase their impact. This book, therefore, brings a sociological approach to power and deviance that widens the lens from straightforward questions of crime and law enforcement. This empirical range will be international in scope, both in terms of global dimensions and more localized dimensions.

Up–to–date research – There are many interesting general works in the field, but all of these are dated. This book brings things up to date.

Theoretical framework – The work offers a new structured approach to engaging the question of whether the digital makes a difference. The examination of multiple, potential affordances across the field of multiple, potential harms offers a more systematic approach than has yet been achieved. Hate (and its subsections of political and interpersonal violence and intimidation), obscenity (in relation to adults and to children), corruptions of citizenship (in terms of being informed/misinformed and being informed upon) and appropriation (by fraud, robbery or piracy) may be afforded more readily by increasing access, concealment and evasion or by incitement, or by any manner of combination. Increased risk by some affordances in one field does not mean risk may not be reduced in another, or that affordances can be balanced between increased scope for harm being set against *other* affordances that increase scope for detection and/or prevention. The range of content, up–to–date research and structure of this work allows a more robust answer to the research question. In so doing, am I working within an existing 'theoretical framework'? The answer is 'yes and no', as this work will draw upon a range of frameworks: actor network theory's notion of affordances is perhaps the most obvious, but I do not think this conception is sufficient. I draw on feminism, Manuel Castells' morphogenetic structural approach (something with an element of a Marxist character – but also tempered with a Weberian streak), and strands of the sociology of scientific knowledge traditions.

Accessibility – While this work sets out to offer an original contribution to knowledge, it also aims to be an accessible work that will engage readers who are new to the field and so be a useful core work for students studying a range of modules at undergraduate (and postgraduate taught) level in the field of cyber crime and society (in sociology, law, criminology and business). To this end, the work will contain three sets of features that will allow a novice to navigate the field:

1. Each chapter will begin with a set of orienting questions that will allow the reader to grasp the key issues being debated in the area under discussion in that chapter. These questions will allow those new to the field to ask themselves what they think before the chapter begins, and they may wish to return to the questions at the end to ask whether their thinking has changed. These questions may act as prompts for discussion in seminars if a tutor chooses to use the book as a guide for teaching.
2. After the orienting questions have been posed at the start of each chapter, there will be a brief 'synopsis' subsection, outlining key disputes and contested claims in the area under discussion. This 'head to head' section will reference the contemporary readings in the field that are generating the most 'heat' in the academic debate. For anyone keen to use the work as a core text for teaching the field of cyber crime, these key readings will be a useful guide to further sources they might suggest students consult after and in addition to the work.
3. Within each chapter there will be detailed discussion of case studies – research that has been of particular significance in the field in question. Again, these case studies, while located in the work as fundamental to the development of its original argument, also act to guide the novice reader.

A note on reflexive epistemological diversity

This work applies to the field of digital network technologies and crime the 'reflexive epistemological diversity' approach first outlined in relation to genetics and computer science in my book *Science in Society* (2005). In that work, four distinct sociological traditions are set out in productive tension with one another. These traditions are: 1. Ethnographic/constructivist approaches (including the actor network theory [ANT] approach); 2. the older 'internal social interests' approach to science and technology studies (STS); 3. feminist approaches to digital networks and harm; and 4. the neo-Marxist (morphogenetic) approach to the network society developed by Castells. These are partial but productive programmes of research, and additional insights arise when each is set in relation to each of the others. While most often dismissive of one another, each programme of research offers insights that remain obscured when looking through the lenses of the other approaches.

The extent to which technology is shaped by economic interests, gender or the internally self-generated interests of networks of developers and users remains disputed; this is even as all three of these are challenged by accounts that focus attention on the reverse, that is, the capacity of technical assemblages to 'act' in relation to the construction of interests and identities by (or for) human agents. The tension in Castells' approach, between the technical mode of development and the (still) capitalist mode of production, can be said to parallel the tension in actor network theory, between technical assemblages influencing the configuration of human interests and identities, and human actions shaping technological development and application. Feminist researchers, likewise, identify both the scope for gender relations to shape technological benefits and harms, but also the scope for new technical systems to reinforce or challenge such social relations.

PART I

Hate

Terrorism and Hate Crime: From the Long Fuse to Hate Speech

Key questions

1. Is hacking infrastructure or digital denial of service best described as terrorism, and how best is it possible to define cyber terrorism in relation to and in distinction from digital hacktivism?
2. How far have digital networks and terrorist networks mirrored one another?
3. In what ways, and with what limitations, can digital networks enable terrorists?
4. Is the internet an unregulated space for terrorist propaganda and hate speech, and to what extent does such content radicalize and/or incite real violence?
5. Have digital affordances been enacted symmetrically, or asymmetrically in conflicts between states and violent non-state political actors?

Links to affordances

Regarding cyber terrorism, the question of access has shifted over time, from delivering physical harm (such as in detonating a bomb or crashing an aeroplane by remote online access) to the delivery of money, propaganda and communications. The capacity to conceal content and to evade authorities exists across global networks, but terrorist networks, and those uploading hate speech in remote jurisdictions, are not as immune from authorities as was imagined 20 years ago. The dark web is not as dark as it was imagined once to be: virtual proxy networks (VPNs) and TORs increase privacy, and can even be combined; garlic routers add even more security. However, entry and exit point vulnerabilities mean such systems are not inviolable. Simplistic accounts of online radicalization, or the idea that Facebook incited the Arab Spring, are misleading, even if incitement to violence online has inspired a small number of lone-wolf actions. The association between digital networks and empowered small actors, allowing them to take on powerful states through asymmetrical (terrorist) tactics, has been replaced in more recent years with a return to inter-state conflict, now by digital means.

Synopsis

In an information society, can jaw-jaw really be a form of war-war? Certainly, the ground has been sown for a clash of liberties: when the right of expression confronts the right to security. This chapter applies Wall's transformation test (see Chapter 1) to the themes of so-called cyber terrorism and hate speech online. Has the digital made a difference? Clearly, digital networks can be used to distribute propaganda by those advocating politically motivated violence, and to spread hate speech; but do such affordances actually translate into the increased violence some seek to achieve? Digital networks can be used to trigger a bomb, but have not been used successfully to crash a plane or blow up a nuclear power station (as has been depicted in fiction). The internet can be used to recruit and plan violence, but such channels are not so anonymous and unregulated as might be imagined.

Chapter sections

1. Cyber terrorism before 9/11: Denning (2000), and the finding that there was little evidence of threat, versus Virilio's (2000) claims regarding 'The information bomb'. Different conceptions of the problem produced very different conclusions.
2. After 9/11: the figure of al-Qaeda as the ultimate network terror organization in the networked age – and the new myth of the distributed network actor (Galloway, 2005).
3. The long fuse, or not? Brenner (2007) on the absence of the online bomber, but other functions come to the fore.
4. The Arab Spring becomes the Syrian Winter, from mass organization to beheading videos (Friis, 2015; Awan, 2017); and far-right hate speech in the US/Europe (Bleich, 2011). Propaganda and hate crime take centre stage. How to distinguish cyber terrorism from digital hacktivism?
5. From empowered small actors to the rise of cyber warfare between states, and state-sponsored cyber terrorism (Singer and Friedman, 2014).

1. Cyber terrorism before 9/11

Looking back on what arose in the 1990s, Michael Stohl suggests:

Over the past two decades there has developed a voluminous literature on the problem of cyber terrorism. The themes developed by those writing on cyber terrorism appear to spring from the titles of Tom Clancy's fiction, such as *Clear and Present Danger*, *The Sum of All Fears* and *Breaking Point*, or somewhat more cynically, *Patriot Games*. (2007, p 223)

These exaggerated representations have led to what Gabriel Weimann (2005, 2015) calls 'cyber angst' (see also Conway, 2011, on media 'hypers' and audience psychology regarding fear of technology and terrorism as 'unknowns'; and Taliharm, 2010, on definitional slippage).

Myriam Dunn Cavalty critically notes misdirection of anxiety over computer-based terrorist threats, citing a 1991 US National Academy of Sciences report as stating: 'We are at risk. Increasingly, America depends on computers ... Tomorrow's terrorist may be able to do more damage with a keyboard than with a bomb' (cited in Cavalty, 2008, p 19). Prognoses of doom here were not based on diagnostic evidence of such threats. The 1990s saw the emergence of greater concern over the possibility of computer-related terrorism (what would come to be referred to as cyber terrorism) precisely at the moment when the internet moved out from a primarily defence industries (and related) research-based network to being the civilian communications infrastructure that it is today. Ironically, perhaps, it was at the point when the internet ceased to be principally an instrument of military communications and coordination that the 'risk' of its use for military communications and coordination by those other than its originators began to be posed as a serious consideration.

In this context, Frederick Kittler's (1997) account of the rise of 'post-war' provided a revised account of computer-based and information-based warfare's history. His (pre-)history of today's information warfare highlights how radio, with its broadcast capacity, was initially rejected by the British military in the 1920s, even when radio could be used to coordinate military actions. This was because enemies could intercept such broadcasts, such that any coordinated action would be known in advance. Kittler notes that German military strategists simply devised ways of encrypting the military orders they subsequently broadcast. Decryption, by means of computers, then became quite literally the 'key' to winning radio wars, creating what Kittler calls 'post-war' (which today might be called packet-switching war). Nonetheless, in such warfare, 'information' was still the means of coordinating and/or finding out about kinetic forms of violent action, not the means of directly delivering violence itself.

Paul Virilio's (2000) *The Information Bomb* develops this construction of the supposed new threat of 'the logic bomb', here proposing two dimensions of the supposed threat of cyber conflict: 'the flood' and 'the crash'. The flood,

Virilio suggests, represents the threat networked computers pose in terms of a world where boundaries cannot be defended and where information flows overload any attempt to limit propaganda and control being exercised by any actor capable of controlling or even accessing the network – leading to new forms of cyber colonization/globalitarianism. Virilio writes:

> After the first bomb, the atom bomb, which was capable of using the energy of radioactivity to smash matter, the spectre of a second bomb is looming at the end of this millennium. This is the information bomb, capable of using the interactivity of information to wreck the peace between nations. (2000, p 63)

Whether this be propaganda or hostile actors taking over critical infrastructure, such a threat, he argues, calls for 'circuit breakers' to limit cultural, political and business 'flooding'. However, even with circuit breakers (in terms of critical infrastructure not being directly linked to the web, or in terms of blocking certain kinds of content), whether or not the flood can be prevented is an ongoing question.

Regarding 'the crash', Virilio (2000, p 134) writes: 'If interactivity is to information what radioactivity is to energy, then we are confronted with the fearsome emergence of the "Accident to end all accidents", an accident which is no longer local and precisely situated, but global and generalized.' The virtue of a system that is globally integrated becomes its own Achilles' heel, as what connects everything can be hacked from anywhere, and the consequence of such damage might then not just spread anywhere but might cause a generalized problem that undermines everything. The threat of the flood (the inability to stop the flow of harmful information) and the crash (the danger of remote actors causing catastrophic damage via digital networks) have become, then, the twin fears regarding cyber terrorism.

What, then, of the substance of such threats in those early years of concern about cyber terrorism? Dorothy Denning (2000, p 6) notes predictions of planes being digitally hijacked and crashed, or nuclear power stations being remotely sent into meltdown, and writes: 'To the best of my knowledge, no [digital] attack so far has led to violence or injury to persons, although some [digital intrusions] may have intimidated their victims.' Denning argues that using digital networks to coordinate, collect intelligence, propagandize, recruit and fundraise should not be called 'cyber terrorism'; instead, she reserves the term to refer to politically motivated 'attacks and threats of attack against computers, networks, and the information stored thereon' that 'result in violence against persons or property, or at least cause enough harm to generate fear' (2000, p 1). This narrow conception remains disputed among researchers, with no overall agreement emerging over time (Jarvis, McDonald and Nouri, 2014; Jarvis and MacDonald, 2015). While Denning

(2000) notes some evidence of some terrorist groups having sought to gain the skills and resources to conduct such attacks, none had succeeded by that time. However, Denning concludes by suggesting that, despite a number of limitations on the scope and attractiveness of network-based violence to politically motivated, non-state (terrorist) actors, the potential to deliver high levels of harm at a distance via digital networks would likely attract terrorists to using the internet as a means of conducting violent acts. She notes that action at a distance (what is sometimes referred to as 'the long fuse') makes 'the logic bomb' (cyber terrorism) attractive relative to 'the truck bomb' (as used by Timothy McVeigh – who, at the time of Denning writing, was being prosecuted for the then largest domestic terrorist act on US soil, the 1995 Oklahoma City bombing in which 168 people died). The attacks on 11 September 2001 (or 9/11) radically altered perceptions on this question. It seemed that despite the supposed advantages of action at a distance, and perhaps precisely in an age of mediated interaction, the direct act – exemplified by the defining 'signal event' of the 21st century, the crashing of passenger jets into the Twin Towers of New York's World Trade Centre – suggests that cyber terrorism was not going to replace the direct act. Mike Davis' (2007) history of the truck bomb concludes by suggesting that it is by the very nature of its technical simplicity that the truck bomb (and, by extension, any vehicle) threatens the image of separation of powerful and powerless in society, that is, it is the 'poor man's airforce': 'A complex weapon makes the strong stronger, while a simple weapon – so long as there is no answer to it – gives claws to the weak' (George Orwell, cited in Davis, 2007, p 4). After 9/11, debate shifted from the logic bomb, to the extent to which the networked terrorist organization, not the networked terrorist act, was key to the significance of the digital in its relationship to terrorism. In 2007, Denning (2007) also concluded that the threat of cyber terrorism, which she had suggested was not imminent (but only potential) in 2000, remained non-imminent in 2007, and that organization by concealed and remote (evasive) communication, rather than to the triggering of a destructive act remained the key benefits of the internet for terrorists.

2. After 9/11

After 9/11, the figure of al-Qaeda came to be seen as the ultimate network terror organization in the networked age – the new myth of the distributed network actor (Galloway, 2005; see also Holt, 2012). Given media tendencies to scale up threats in terms of actor competence, target and action (Jarvis, MacDonald and Whiting, 2016a, 2016b), misconception easily arises.

As mentioned earlier, just prior to 11 September 2001, Dorothy Denning (2000) suggested that while the internet was a possible asset when it came to (evasive, concealed and remotely accessing) organization, as well as propaganda and recruitment (incitement), the use of digital networks by terrorists should not be considered 'cyber terrorism', unless such actions were directed against digital infrastructure with the effect of producing real-world physical harms to people or property in pursuit of a political objective. Denning's suggestion in 2000, that the logic bomb might replace the truck bomb in the future, however, seemed wholly refuted on 11 September 2001. Replacing trucks with aircraft appeared to take the logic of the truck bomb to a new level, rather than replacing it with the logic bomb's principle of action at a distance.

The response to 9/11 was to shift attention away in conceptualizing cyber terrorism, from 'the long fuse' (the logic bomb) towards the question of terrorist organization. If 'access' (triggering remotely by a long fuse) was no longer the aspect of the digital that was seeming to make a difference, then perhaps it was the digital capacity to evade and conceal terrorist organization that was key. The shift in attention from kinetic access to organizational access (itself also concealed and evasively distributed) by terrorist organization was most fully articulated by New York University professor Alexander Galloway, in his account of the contrast between the World Trade Centre in New York (towers symbolizing hierarchical power) and al-Qaeda (the ideal-typical networked terrorist organization).

Galloway (2005) notes that the internet was developed as a distributed network system. Information is broken down and routed to its destination via 'packet switching', such that any blocked pathway is 're-routed'. The basic design is that no node in the network is essential as any blockage can be re-routed. This system was designed to avoid communication breakdown in the Cold War scenario where a nuclear attack might disable numerous nodes in US defence communications. Such a distributed architecture was designed to fight the centralized Soviet military machine, but has, today, created a network architecture that can itself be exploited by new networked actors able to take advantage of such ability to evade and conceal online.

Galloway suggests that the group responsible for organizing the 9/11 terrorist attacks in 2001, al-Qaeda, precisely represent a viral organization, ideally adapted to evasion and concealment in a new networked age. 'Because the Internet is globally interconnected, a single virus will likely have massive repercussions, because the Internet is so robust, viruses can route around problems and stoppages. And because the Internet is so decentralised, it is virtually impossible to kill viruses once they are released' (Galloway, 2005, p 27). The cell structure of the organization is precisely designed to enable autonomous action and minimal traceability. While the attacks on 11 September 2001 extended the kinetic logic of the truck bomb, rather than the long-fuse affordances of the logic bomb, taking the mechanism of the suicide attack to a new level of violent extremes, the organizational principles of al-Qaeda appeared to demonstrate a new level of distributed power relative to the vulnerability of traditional hierarchical power – as manifested in the Twin Towers. The New York skyline, perhaps the most iconic manifestation of the power of 20th-century global capitalist civilization, seemed suddenly to appear a manifestation of inherent insecurity in an age of distributed network actors:

> All the words used to describe the World Trade Centre after the attacks of September 11, 2001, revealed its design vulnerabilities vis-à-vis terrorists: It was a tower, a centre, an icon, a pillar, a hub. Conversely, terrorists are always described with a different vocabulary: They are cellular, networked, modular, and nimble. Groups like al-Qaeda specifically promote a modular, distributed structure based on small autonomous groups. They write that new recruits 'should not know one another,' and that training sessions should be limited to '7–10 individuals.' They describe their security strategies as 'creative' and 'flexible'. (Galloway, 2005, p 28)

However, the history of such asymmetric warfare, Galloway argues, goes back at least as far as the American War of Independence. It is simply necessary to adapt. Galloway points out that the protocols upon which the internet depends, and which machines are required to adopt in order to be able to communicate, remain relatively centrally controlled. Power within distributed systems has not, therefore, disappeared, but has, rather, shifted to control over protocols – something that still remains largely in the hands of a small network of US-based computer scientists (Galloway, 2004, 2005).

Galloway notes that US military power to invade and occupy a territory like Afghanistan is limited when set against the distributed networks of local opposition on the ground. The sweeping invasion of that country in 2001, just like the similarly rapid invasion and occupation of Iraq two years later, was notable for its speed and for the way massed military opposition melted away. However, such opposition did not disappear, but, rather, operated a networked resistance, which in time wore down the invading armies. However, while 'boots on the ground' could not hold a territory against a determined local network, it is also true that, online, organizations like al-Qaeda have been radically degraded and dismantled by state actors that have adapted to the nature of power in distributed environments. In this vein, David Benson (2014, p 293) rejects the view that the internet is a force multiplier for terrorists: 'Far from being at a disadvantage on the Internet, state security organs actually gain at least as much utility from the Internet as terrorist groups do, meaning that at worst the Internet leaves the state in the same position vis-à-vis terrorist campaigns as it was prior to the Internet.' When Ayn Embar-Seddon (2003) argued that the internet was a force multiplier for terrorists, this was based on the observation that many elements of critical informational infrastructure were directly linked to the public internet in the early 2000s. Between then and when Benson was writing, far greater levels of security have been put in place to reduce this. However, in 2015, Abdel Bari Atwan (2015) suggested the lightning speed with which ISIS (Islamic State in Iraq and Syria) could expand into what he called its 'digital caliphate' was in no small measure driven by its capacity to recruit, communicate anonymously and propagandize online. Yet, as will be suggested in later sections of this chapter, the rapid collapse of this caliphate may also show the limits of the digital in securing terrorist ambitions.

Manuel Castells (1996) argues that one dimension of the collapse of the Soviet Union was its inability to adapt to or adopt the affordances of the new networked mode of development. One clear manifestation of this was the defeat of the Red Army in Afghanistan. Soviet tanks (the tank being emblematic of Soviet communism, of its triumph in World War II and of the regime's capacity to occupy Eastern Europe until 1989) were no match for computer-guided missiles supplied to Afghan rebels by the United States. In 2001, it appeared that the tower, iconic representation of the centralized and hierarchical corporation, might go the way of the tank, when faced with the new power of distributed digital networks. This has not happened. While digital networks do afford new forms of organization that can evade and conceal, such affordances are limited by the capacity of states and corporations to reorganize, as they have typically done in a way that the Soviet Union did not. As Section 4 points out, however, even if concealment and evasion have their limits, the capacity to incite may remain – despite the fact that the rise of the lone wolf may, in such circumstances, replace the 'cell' structure of a still centrally controlled organization like al-Qaeda. Cynthia and Michael Stohl (2007) point out that digital networks are, in fact, far more distributed than terrorist networks. Terrorist networks tend to have forms of centralized power, like states, that digital network surveillance can be used to identify. The power of virtual warfare may, then, be in creating external threats and then incorporating them, even if such virtualization strategies may be subject to disruptions (Lundborg, 2016). Terrorist networks are not as distributed (concealed and evading) as Galloway suggested; and states have deployed more advanced and distributed network technologies against them.

3. The long fuse, or not?

The absence of the online bomber seemed to discredit earlier conceptions of cyber terrorism, but other digital affordances come to the fore – in terms of recruitment and coordination (Lachow, 2009), fundraising and money laundering (Keene, 2011; Salami, 2017), propaganda (Choi, Lee and Cadigan, 2018) and attempts to hack infrastructure (Jarvis, MacDonald and Chen, 2015). Susan Brenner (2007) argues that the mode by which a crime is committed should not define the nature of the crime; rather, it should be the motive of the actor and the nature of the target that does so. This leads her to argue that defining crimes as 'cyber crimes' simply because they have involved some degree of digital networking to organize the crime is misleading. She argues it is no more useful to refer to the 11 September 2001 terrorist attacks as 'aeroplane crimes' as it would be to refer to a bank robbery as an 'automotive crime' if a getaway car was involved in the conduct of the robbery. Limiting the category of 'cyber terrorism' to the actual conduct of acts leading to physical harm by digital means, Brenner concludes: 'To date, there have been no known instances of cyberterrorism' (2007, p 389). Prior cases – where, for example, computers were hacked to release raw sewerage, and to disrupt air-traffic control systems – were not, in fact, politically motivated, Brenner argues. Even if they had been, Brenner questions whether such harm as might have arisen (that is, poisoning or from an aeroplane accident) might best not be called 'cyber terrorism' if it was the sewerage or the crash that caused the harm, not the computer. Nevertheless, despite Brenner's wish to maintain a very narrow conception of what 'cyber terrorism' would be, she does go on to outline a three-dimensional model of how computer networks may facilitate terrorism, regardless of what this might then be called: weapons of mass destruction, weapons of mass distraction and weapons of mass disruption.

'Weapons of mass destruction' refers to the means by which violent acts may be carried out, in this case by digital means. While, in 2007 at least, Brenner did not believe such an act had been carried out for the purposes of supporting a non-state actor in the conduct of a politically motivated campaign, she suggests such acts could happen in the future. The example she gives is the act of hacking into a nuclear power station to create a meltdown, such as the one that occurred at Chernobyl in 1986. The meltdown, if terrorists could successfully claim responsibility, would undermine the credibility of the regime they opposed, such as was the case with the Soviet Union after 1986; the meltdown at Chernobyl was not, however, the result of terrorist hackers. Whether or not failures in Iranian nuclear reactor centrifuges caused by the US/Israeli Stuxnet virus (Kenney, 2015) should be called cyber terrorism or cyber warfare is a more complex question, both because the hackers were state actors and because nobody was killed. To be able to take credit for such an act might be highly desirable for a terrorist group to achieve. Yet, Brenner (2007, p 391) concludes: 'To describe this scenario as cyber terrorism is as inappropriate as describing the 1998 U.S. embassy bombings carried out by al-Qaeda as automotive-terrorism because vehicles were used to deliver the bombs to the target sites.' Further, using David Wall's (2007) transformation test, it must be asked: if the suicide bombers in those attacks had not had motorized vehicles and had to attack their target on foot, would the acts have been as destructive as they were? If not, it is reasonable to say that these were, in some sense at least, 'automotive' crimes, as without vehicles the crime would have been stopped or radically diminished. Nevertheless, regarding cyber terrorism, Brenner is certainly correct to say that, relative to the kinetically delivered suicide attack, the digital delivery mechanism of a nuclear reactor meltdown, aeroplane accidents or mass poisonings remain a possibility but not a substantive reality.

'Weapons of mass distraction' (inciting fear and confusion) are, for Brenner, a much more credible threat relative to the affordances of digital network communications. In this scenario, a kinetic act of violent destruction that was not 'cyber terrorism' would be amplified by the delivery of propaganda online – what Brenner (2007, p 397) refers to as 'a terror multiplier'. Psychological manipulation of the civilian population is key to what defines 'terrorism' as 'terrorism', that is, the attempt to instil 'terror', not just the conduct of violent acts as such. Circulation of propaganda online may act to intimidate civilians, discredit a target regime, radicalize potential followers and even to recruit new members. Civil unrest might ensue, with attendant destruction and violence. The morale of state actors might also be undermined, such as when military opposition to a terrorist group may crumble in the face of the perceived levels of violence that soldiers (often conscripted) might believe they are confronting. The British army has, for centuries, used the Nepalese Ghurkha regiment's curved Kukri knife as a psychological weapon to encourage opponents to surrender rather than face the prospect of being beheaded in the night by Ghurkha soldiers who are sent out to raid enemy positions. That ISIS in Syria and Iraq used similar threats of beheading, to encourage opponents to abandon their positions and retreat before confronting ISIS fighters, gained added impact by the online distribution of filmed beheadings (locally, at least). Whether such psychological warfare was successful in the end is debatable (as will be discussed in the next section). Brenner notes that in the 1930s, accounts of 'mass hysteria' associated with Orson Welles' radio theatre company's broadcast of H.G. Wells' *War of the Worlds* were highly exaggerated. Whether the power of online propaganda for recruitment, radicalization and intimidation is similarly exaggerated today is debatable. That digital networks have been used to engage in such tactics cannot, however, be disputed.

Finally, 'weapons of mass disruption' (access to critical infrastructure) refers to when the 'terrorists' goal is the infliction of system damage on one or more target systems' (Brenner, 2007, p 393) – so, using digital networks to take down digitally controlled state and corporate infrastructure: financial, medical, energy grids and so on. The increasingly computer-controlled nature of such infrastructures does create scope for attack, and there is certainly a cat-and-mouse struggle between such networks and those that would seek to identify and exploit their connectivity. In 2007, Estonia saw a politically motivated attack by Russian patriot hackers against its largely internet-based media and government communications infrastructure. This act of cyber warfare may or may not be best described as cyber terrorism. It was not an act of direct violence, but it was designed to harm and intimidate. One thing this attack did do, however, was to ensure Estonia and other states immediately sought to add firewalls between essential digital infrastructure and the public internet (Herzog, 2011). Now, such systems are not routinely run or connected directly to public digital networks. However, connections can be located and exploited. The impact of COVID-19 saw many switch to working from home, and, as such, connection to critical infrastructure via home computer networks was introduced as an emergency measure, creating a vulnerability that was exploited by some hostile actors. The nature of such actors brings us to Brenner's final, and most substantial discussion, of who these actors are, and of how best to tell.

For Brenner, while digital networks do create potential vulnerabilities that could be exploited by terrorists, the real problem with digital networks is that they make it harder to identify the source and intention of those committing harmful acts. The distinction between internal law and external war is disrupted when digital networks flow around and across borders; and, similarly, the distinction between state actors engaged in cyber warfare and non-state actors engaged in cyber terrorism is disrupted when it is harder to locate combatants and when their physical location may not be proof of whether they are criminals, terrorists or state-sponsored agents. Stan Gilmour (2014), likewise, suggests the biggest problem has been in deciding which 'policing' agents should take charge, coordinate and train accordingly (service providers, website managers, corporations, formal police or military services). A consequence of Brenner's concern here has been a rapid escalation of investment in security measures; and, as is discussed in Section 5, the asymmetry of state and non-state actors has, in large measure, been reasserted accordingly (digital affordances do not change outcomes if currently dominant actors – states – remain better able at deploying them). Access by authorities may undo terrorist network potential to evade and conceal.

4. Incitement and hate speech

As the 2010s developed, the Arab Spring became the Syrian Winter. Hope for mass-organized opposition, inspired by online messaging (such as Castells' 2012 account of online organizations in Egypt in overthrowing the Mubarak regime), moved to a more pessimistic fixation upon beheading videos (Friis, 2015; Awan, 2017). Far-right hate speech in the US/Europe also spread online (Bleich, 2011). Propaganda and hate crime took centre stage. The question of how to distinguish cyber terrorism from digital hacktivism also arose in this context.

As Sections 1–3 have highlighted, the concept of cyber terrorism has changed over time. 'The long fuse' itself defused after 9/11. The idea of the networked terrorist organization al-Qaeda being the unstoppable manifestation of a new cyber-terrorist threat also rose and then fell away. Then the idea of 'weapons of mass disruption' came to the fore, but this phenomenon does not appear to have manifested in the way many had feared and/or predicted (perhaps precisely because such fears/predictions led to a sufficient level of protective measures to prevent such threats actually being made real). In more recent years, the digital capacity to incite politically motivated violence has come to the fore.

The rise of photography in the 19th century saw concerns over explicit sexual images rise. Pornography was perceived as far more of a problem than hitherto when it became available to a mass audience. Fear of new-media technologies spreading obscene materials is not new – from the printing press in 16th-century Europe, to cinema, radio, television and video cassettes in the 20th century. Today, the internet distributes obscene material, and, in the case of 'cyber terrorism', it is suggested that hateful speech and violent images that glorify and promote terrorist violence will themselves inspire imitation and hence increase terrorism. Far-right racist Anders Behring Breivik sought to distribute an online manifesto justifying his 2011 mass murder in Norway; meanwhile, in 2019, the murderer of 51 people in twin attacks on Muslims in Christchurch, New Zealand posted a manifesto online, seeking to justify his actions, and livestreamed these actions in the hope of inspiring others to follow his example. The online distribution of beheading videos by ISIS members in Syria and Iraq in the 2010s was, likewise, accompanied by online calls for viewers to carry out their own 'lone-wolf attacks' in Western countries.

It is commonly believed that the United States allows free speech in a way that is not tolerated in other countries, and that as the initiator of the internet, these principles of freedom of speech are extended worldwide by means of digital networks. As such, it is claimed, the internet is a playground for hateful propaganda and for the incitement to (not only political) violence. While the United States' first amendment does protect the right to free speech, it should be noted that incitement to cause violence is a crime in the United States; and the US Hate Crime Prevention Act 2009 extends protection from hate crime to a wide range of groups (Bleich, 2011). However, this law relates to the conduct of crimes where prejudice against an identifiable group compounds the crime, rather than making the expression of prejudice a specific crime on its own. As such, it is a crime to assault someone or to threaten to do so, or to call on others to do so (that is, incitement). To do any of these things, while having or making reference to that person's membership of an identifiable social group (race, religion, nationality, sex, sexuality, gender or disability, for example) would, then, warrant additional criminal sanction under the 2009 Act. The United States has, therefore, passed substantial new hate crime legislation – in large part as a response to the rise of online communications.

What makes the United States different from most other countries is the distinction between hate crime (as defined in the previous paragraph) and the criminalization of hate speech, where diminishing and dehumanizing a group of people becomes a crime in itself, even if such insults are not directly linked to any other crime or incitement of others to commit such an additional crime. Bleich (2011, pp 928–929) notes the common argument, which is that Europe (and, in particular, Germany), having experienced the Nazi holocaust, was more willing to prohibit 'hate speech' (as distinct from 'hate crime') than was the United States. However, Bleich points out that while Austria introduced prohibitions against Nazi speech and symbols in the 1940s, Germany, France and other European countries did not do so until the 1990s or later, precisely at the point when the internet appeared to threaten to make the spread of such speech more pervasive (see also McGonagle, 2010, on European developments).

Anxiety that online propaganda, and obscene images of violence in particular, would 'radicalize' viewers, such that they might be recruited to terrorist groups or imitate such actions, has come to predominate over other supposed affordances of the internet in facilitating terrorism. I take the example of beheading videos here, but the same argument and issues relate to terrorist mass shootings being disseminated online as well. Simone Molin Friis (2015) examined the case of ISIS beheading videos, particularly those of Western hostages in 2014. It was the expressed intention of those making recordings of these killings that their distribution should intimidate Western governments and populations and dissuade them from intervening in the Syrian civil war, rather as the release of similar videos showing the beheading of Syrian and Iraqi government soldiers had successfully reduced the morale of such forces. Friis, however, notes that videos of Western hostages being beheaded were scrubbed from the internet within minutes, or hours at the most, and that only a very small number of civilians would have ever seen the actual decapitations. (See also Akdeniz, 2010, Banks, 2010 and Rosen, 2013 for discussion of the tension between counter-terrorism and free expression when state and non-state actors block content online.) Clips and stills from beheading videos were used in Western media coverage of events. However, the clips were framed as evidence of the need to intervene. Friis documents how coverage of these images, not the unedited content as such, coincided with a substantial shift in US/UK public opinion: from being largely hostile to military intervention in Syria, to a majority being in favour (if only of air strikes rather than ground-based troop deployment). The claim that such 'instant icons' had the power to reframe the narrative proved false, as control over such images remained largely in the hands of mainstream media channels, who selectively framed and interpreted the videos, and of new-media platforms, who immediately removed the full-length originals. In contrast to claims in the mainstream media, that beheading videos illustrated how new media was out of control, Friis concludes that it is precisely this kind of framing that is useful in ensuring such online content remains largely under control. While those making and releasing such videos sought to control the narrative, they failed; meanwhile, Western political and media actors were successful in using 'the depoliticizing effects of reducing a complicated conflict to a fragmented visual icon' (Friis, 2015, p 739). The use of war photographers since the Crimean War in the mid-19th century has been managed by dominant political and media actors to create interpretive frames that favour dominant interests. The suggestion that the internet collapses that capability appears to be false. However, what is true for the majority of viewers is not necessarily the whole story. The balance, between encouraging counter voices to challenge hate speech (Citron and Norton, 2011) and editing search algorithms, labelling sites and blocking at various levels (Cohen-Almagor, 2014), remains contested. Where al-Qaeda followers used Facebook (which then adopted a very aggressive delete policy), ISIS followers tended to use Twitter, who, while deleting accounts that advocated terrorist violence, has defended stronger privacy and free-speech policies that slow down such deletion practices (Greenberg, 2015).

Imran Awan (2017) examined 100 Facebook accounts and 50 Twitter feeds associated with ISIS. This work is mirrored in research on Islamophobia (Awan, 2016) and far-right hate groups (Perry and Olsson, 2009). One aspect of Awan's analysis of ISIS online was to examine reactions to such things as beheading videos. Even among viewers that engaged with those who posted supportive comments on such materials, the vast majority of responses were hostile, the great majority of which came from Muslims condemning such actions. However, even while most responses were negative, some were not. Particular groups were more likely to respond positively, and, of these, Arabic speakers in Western countries appeared to have been the most successfully targeted – even though appeals to such groups were not generally successful, only attracting a small number of positive respondents. Even if reaching a 'global audience' via online channels is possible, most of that audience will remain hostile and largely contained within anti-terrorist mainstream interpretations of what they are seeing or being told about. If the goal were to persuade the majority, such a goal appears to be a failure. It is not the case that audiences are, on mass, conditioned into believing terrorist messaging, even if they are exposed to terrorist propaganda – which, as Friis points out, most are not, even when they are exposed to edited versions of terrorist content. However, if the goal is to target those individuals most likely to be persuaded, and who are already disaffected and alienated from mainstream society, radicalization may occur. Awan (2017, p 140) suggests that a kind of social learning theory explanation may account for how a small number are drawn into echo chambers that reinforce a narrowing of associations, definition, differential reinforcement and, eventually, perhaps even imitation. Even if it is an illusion, a belief that the internet is anonymous may also act to disinhibit and encourage 'deindividualization' within small pockets of people. It is not general persuasion but, rather, polarization that furthers terrorist aims, if the desire is to promote lone-wolf imitators, rather than larger scale political movements or organizations. Encouraging alienated individuals to carry out violent acts of the most 'simplistic' nature (knives, guns or even driving into crowds, rather than anything 'sophisticated' in either technical or organizational terms) may be facilitated by digital networks. Nevertheless, the shift, from encouraging people to join ISIS and to come and fight in Syria/Iraq, to promoting lone-wolf actions in Western countries, was also a reflection of the physical dismantling of ISIS 'in the real world', as their self-declared caliphate was extinguished. Whether a Taliban government in Afghanistan sees a return to such a geographically located threat takes us to the question of state-sponsored cyber terrorism more generally.

5. From small actors to state-directed cyber warfare

With the rise of cyber warfare between states and state-sponsored cyber terrorism, more recent discussions about digital warfare have switched from notions of asymmetric conflict, such as between non-state political actors and states, to conflict between states (Ohin et al, 2015; HM Government, 2016). These state actors are not, however, always evenly matched either. Earlier concerns about non-state terrorist groups being able to use the internet to outmanoeuvre more traditional, bureaucratic state structures (whether military or otherwise) have become less significant precisely because such concerns were taken seriously at the time, and still are to a large degree, meaning now that states have developed the capacity to 'weaponize' the internet (if information gathering, communications and propaganda are best seen as 'weapons', which Brenner disputes). The capacity of states to harness the power of the internet has led to increased discussion of cyber warfare rather than cyber terrorism (or efforts to counter cyber terrorism) (Brown and Kraft, 2009). The capacity of states to engage in 'terrorist' acts (such as targeting civilians) should not be forgotten.

Singer and Friedman (2014, p 96) write: 'Thirty-one thousand three hundred. That's roughly the number of magazine and journal articles written so far that discuss the phenomenon of cyber terrorism. Zero. That's the number of people who had been physically hurt or killed by cyber terrorism at the time this book went to press.' These authors conclude that the more substantive threat comes from information-based conflict between states, rather than from non-state terrorist organizations. While disruption can still be caused by hackers acting alone, or as part of wider 'hacktivist' networks (as we will soon see), whether these actors are motivated by political ideologies, personal grudges or material gain, the scale and scope of states and corporations to meet such challenges has increased precisely because of earlier waves of concern. Earlier waves of concern have seen governments invest heavily in cyber security. Such defensive capabilities have also given rise to the increased capacity of state (and some large corporate) actors to move from defensive to offensive cyber warfare, well beyond the capacity of small-scale actors to match or defend against.

The notion of the hacktivist itself emerged from an earlier conception of the hacker as politically neutral 'Geek', simply seeking to see how things worked and how far they could be altered or accessed without permission. As such, the hacktivist emerged as a politicized hacker, in line with the rise of the public internet in the 1990s. Whether in the form of the mass-action hacktivist, using digital denial-of-service attacks to crash servers or websites (in a fashion, mimicking the traditional mass protest/occupation/picket), or the more individualized strategy of access, defacement or damage, or the 'digitally correct hacktivist', whose goal was to 'set information free', the image of the hacker as anti-establishment rebel came to the fore in the early 2000s (Jordan and Taylor, 2004; Söderburg, 2008). While a number of terrorist organizations sought to use hacking tactics in that period, the association between hacktivists and terrorists is limited, as was the capacity of cyber terrorists relative to those state actors they set themselves against. Today, the efficacy of what are referred to as 'patriot hackers' (hackers employed by states to engage in defensive but also increasingly offensive cyber warfare on behalf of 'their' nation states) far exceeds the scale and capacity of non-state-based political rebels. Hackers have become an intrinsic part of the establishment, or at least of the state apparatuses of almost all nation states.

The rise and fall of Anonymous illustrates the shift in the 2010s, from the image of the powerful anti-establishment hacktivist to an increasingly powerful set of state actors harnessing the potency of hackers as a cyber-warfare tactic against other states in particular. From 2008 to 2011, Anonymous came to prominence through a series of 'hactions' against Scientology, corporations who were seeking to prevent support payments to WikiLeaks after the arrest of its founder Julian Assange, and against a number of digital security contractors. The group's use of the iconic Guy Fawkes mask to present its 'anonymous' face seemed to sum up its claim to be everywhere and, at the same time, untraceable. This image of an untouchable digital vigilante organization was, however, short lived. Within a short period, a raft of supposedly anonymous actors involved in a range of the organization's high-profile hactions had been arrested and jailed in a number of countries. It turned out that state actors, working in isolation and in tandem, were able to identify participants much more easily than the hackers themselves had apparently imagined. In 2013, Edward Snowden's revelations regarding the scale of the United States Government surveillance brought to light just how far states had come in this capacity.

Cyber warfare is, to a large extent, about information collection, whether that be espionage in the most traditional sense of collecting military secrets or in the wider sense of gathering information of economic advantage or of potential use in engaging in political leverage (such as hacking into Hillary Clinton's email account). In an increasingly digital age, information becomes central to geo-political power struggles, whether it be in terms of intellectual property that may secure economic advantage between states, or in terms of the capacity to conduct military action. Most acts of cyber warfare are espionage (Maurushat, 2013; Albahar, 2017). Information collection can morph into more offensive cyber warfare if such information can be used to then disrupt an opponent's capacity to defend themselves, either because, for example, their digital infrastructure can be disrupted or that information reveals the location of more traditional military targets (which could include not just the position of assets in a war but also the names, addresses and family contacts of key personnel). While non-state terrorist actors have used digital technologies to identify targets from simple information leakage (such as might be revealed in what a soldier inadvertently includes in a photograph they post on a personal social media account), this is dwarfed by what state-sponsored hackers can scrape online or from hacking into supposedly more secure sites and servers.

Nonetheless, the power of the state in today's cyber-warfare age is, in part at least, a double-edged sword. Where multiple states seek advantage, one over another and in various alliances and networks of trust and distrust, various technological artefacts have been developed to counter the efforts of other states, and these then become barriers to the unlimited actions of states, both relative to one another and in relation to citizens. During the aforementioned Arab Spring in 2011, attempts by various authoritarian governments, such as Egypt under Mubarak, sought to 'shut down' the internet to limit the communications of protestors. This was itself limited by social media platforms, with the support of the US government, who provided their users in the Middle East with 'workarounds' that allowed continued communication. Likewise, the US government sponsored the development of VPNs and TORs, to enable users of the internet to bypass firewalls imposed by various other states (in particular China). Supposedly anonymous communications channels, like the Russian Telegram messaging service, were designed to allow users to evade US-based surveillance, although whether they are then prone to Russian state intrusion (or those of the United Arab Emirates – where the company is now headquartered) is another matter. In any case, the competition between states does mean no single state has absolute oversight, and so channels of anonymous communication, or relatively secure encryption, remain – at least for limited periods (systems, like the dark web for example, will eventually become accessible given sufficient effort and resources by state-sponsored reverse hackers). While the key elements of hacking – access, revelation and damage to digital systems – are of use to terrorists, the capacity of states to engage in such strategies, both against other states and against non-state actors (including, but far from exclusively, against terrorists), means states and certain corporations have become the dominant player in today's networked world. Some states consider certain information as essential to freedom, while other states see that same information as a threat to national or personal security. Whether it concerns economic information (intellectual property, for example), political information (such as might be critical of a regime) or cultural debate (in terms of free speech versus hate speech), such disputes between states over what citizens should be allowed to know, say and do with certain kinds of information creates space for non-state actors. Some of these non-state actors will have criminal or politically violent intent, but it is very important to remember that the vast majority of such persons are not and should not be seen as terrorists just because they disagree with their governments and/or wish to retain certain privacies from their state. The capacity to conceal is limited but real, and whether what scope there is should be viewed as a threat or as an essential freedom should be for society to decide. It is not determined by technology, nor by any intellectual fiat. It should also be asked whether 'we' are too willing to be terrorized by potential threats, when other more pressing issues cause far more harm every day (Furedi, 2007).

In conclusion

Despite the supposed capacity to either crash or flood digital networks in order to exercise political leverage over societies increasingly dependent upon such infrastructures, the scale of such terrorist action at a distance (an example of access) by digital means remains limited – not least because state actors have developed by far the greater capacity for cyber warfare than have any non-state political actors. That the terrorist signal events of the 21st century have rejected action at a distance for the direct act (suicide bombing and beheading videos) shifts attention from the digital delivery of violence (the network trigger of a bomb or the crashing of a plane) to the digital delivery of its recording (the very opposite of concealment). However, other forms of evasion and concealment are significant (in terms of organization, financing and planning), although, again, the limits of concealment and evasion in an age of digital surveillance and drone strikes does suggest that state cyber warfare remains dominant. That beheading videos and suicide attacks broadcast online have incited others to carry out violent acts is significant, but those radicalized remain a tiny number and radicalization takes far more than simply being 'switched on' by online content. The management of such content means the vast majority of viewers see such content framed within narratives that discredit the acts – narratives that, thus, perhaps reinforce resistance to such acts and actors, rather than encouraging them. Overall, the focus of concerns has shifted from fear of networked access (the long fuse) in the conduct of kinetic acts of violence, to the affordances of concealment and evasion in terrorist organization, and finally to scope to incite violent action by means of remote radicalization and propaganda. In each case, these concerns have proved overblown. This is for various reasons, but in particular the capacity of states to both out-perform terrorist groups in the deployment of digital affordances, and to become increasingly sophisticated in conducting cyber-warfare with each other.

Bullying, Stalking and Trolling

Key questions

1. How far have digital networks changed the demographic characteristics of stalkers and their victims; and to what extent has the risk posed by stalkers increased or declined as networks increase the scope for access at a distance?
2. Are children and young people today more at risk of bullying than previous generations?
3. Does the disappearance of disappearance, the increased visibility of people in online environments, increase the scope for victimization, or the scope for authorities to identify bullies, stalkers and trolls? Can both things be true at the same time?
4. Is the online realm more threatening because it is so all-embracing, or less threatening than real life as it is always possible to find your tribe somewhere else online even when you cannot always do so in your immediate physical surroundings?
5. In relation to online trolling, are we simply less polite to one another today, or rather have we become more polite, and hence less tolerant of people who can access us, even as we are by the same means incited to access them back?

Links to affordances

Access to victims by digital means is what defines cyber bullying, stalking and trolling. While digital networks do, without doubt, increase access to victims in the sense of communications, this is not 'access' as unauthorized and unwanted 'intrusion' in the physical sense, or physical violence akin to forms of face-to-face, physical bullying. As such, access may be both increased and reduced by online means, although forms of online surveillance and self-revelation may increase the scope for physical access. Concealment and evasion can be enhanced by online means, but the disappearance of disappearance, along with various forms of online and real-world regulation, means that revelation and punishment can follow. The capacity of digital harassment to incite harm is real, but victims can fight back and use networks to find more fulfilling interactions.

Synopsis

The changing composition of stalkers and the new dynamics of interpersonal revelation, surveillance and interaction suggest that the digital really has made a difference. However, while stalking online shows different characteristics from that carried out in predominantly 'real-world' space and time, the most harmful forms of stalking and bullying retain many traditional characteristics, which trolling now replicates as well.

Online interaction increases scope for being targeted, and expectations to reveal information about one's actions, location and preferences creates opportunities for bullies, stalkers and trolls. Can we and should we seek to prohibit certain forms of display and surveillance, and does that mean protecting some people from themselves (stopping certain kinds of interaction/display/revelation) or blaming internet users for their own victimization?

While particular spaces swarm with trolls, finding one's tribe is made easier by online networks, so, while 'mobile network youth' (Castells et al, 2007) experience new forms of harassment online, it is also increasingly possible to find communities of support and identification, more than was and is possible in real-world interactions.

Chapter sections

1. Key terminology changing meaning over time. Terms like stalking and bullying have changed their meaning over time (Owens, 2016), as has the kind of action that would be seen as stalking or as bullying. Terms like trolling and catfishing have emerged.
2. The disappearance of disappearance. Online access and concealment have to be set against scope for identification by means of 'digital mouse droppings'. Digilantism – digital vigilantism (Jane, 2016) – is one mode of rendering networks 'symmetrical', although not one without significant problems. The relationship between surveillance and concealment/evasion is complex (Mason and Magnet, 2012).
3. Stalking has changed but, in some ways, remains the same. On the one hand, today's 'mobile network youth' experience higher levels of stalking-related intrusions than did older generations (Berry and Bainbridge, 2017), in something that has been referred to as both the normalization and the democratization of stalking. However, the most dangerous forms

of digital stalking retain the same gendered and relationship patterns of victim and perpetrator that existed (and exist) in physical-world stalking (Southworth et al, 2007).

4. Children that experience physical bullying are also the most likely to be bullied online. For some, the digital can be more invasive as it reaches into the victim's private space, while, for others, the digital offers scope for escape and for finding like-minded others not available in the victim's physical environment. While some researchers focus on cyber bullying in children (Mishna et al, 2012), others identify the rise of adult-to-adult bullying online (Nycyk, 2015).

5. Trolling of feminists (such as in Gamergate; see Nagle, 2017) highlights the crossover between interpersonal bullying and group-directed (and, hence, political) hate speech (Vera-Gray, 2017). Trolls are able to access victims remotely with a degree of concealment and evasion of regulation; but network communities and wider policing actors online are able to regulate community standards in many cases to exclude, expose and challenge trolls.

1. Old terms change meanings and new terms/intrusions arise

Stalking is the persistent, unwarranted and unwanted intrusion into another person's life. This may be by digital means, but prior to the rise of the internet, stalking by physical intrusion or by persistent communication saw the term come into common use and, since the 1990s, form the focus of legislation. 'Cyber stalking' behaviour can be part of an overall pattern of behaviour that includes 'real-world' stalking, but most online stalking behaviour neither extends from or transfers into 'real-world' contact/intrusion. The rise of cyber culture has also altered the way language is used in relation to stalking, even as some forms of behaviour remain largely unchanged. In a study of coercive and abusive relationships, Woodlock (2017, p 584) identifies how digital affordances can be used to facilitate control: 'Technology was used to create a sense of the perpetrator's omnipresence, and to isolate, punish, and humiliate domestic violence victims. Perpetrators also threatened to share sexualized content online to humiliate victims.'

Before the rise of the internet, most behaviour defined as 'stalking' was conducted by male ex-intimates and was directed towards female ex-intimates. Around three-quarters of such interactions defined as instances of stalking were of this kind (Roberts and Dziegielewski, 1996; Owens, 2016). A strong link exists between women reporting a male ex-intimate for physical stalking and abusive behaviour during the course of the time that the victim had been in an intimate relationship with their subsequent stalker (Tjaden, 1997). Where no prior violence had occurred, subsequent contact between the ex-intimates is less likely to be perceived as threatening, and hence less likely to be reported to the police. Ngo (2018) found that the police were far less likely to recommend further investigation or to pursue prosecution where a victim of stalking did not believe the perpetrator was an ex-intimate. While real-world stalker ex-intimates do appear to be the most significant threat, in particular to female victims, there is a tendency for police to be reluctant to take cyber stalking reports (whether they concern an ex-intimate or another party) as seriously as they do real-world intrusions, which may sometimes prove mistaken in retrospect.

The rise of online communication has seen a significant increase in the incidence of people receiving persistent, unwarranted, unsolicited and unwanted communications – and such communication has come to be seen as a new form of stalking. Such intrusion by online means does not involve 'traditional' intrusion in the sense of physical invasion of private spaces, but then neither did persistent telephone calling in past times. What is most significant about such online stalking behaviour is that most of it does not come from ex-intimates. While in real-world stalking there is a significant gender difference in the pattern of victimization, women are only slightly more likely than men to say they have been victims of stalking online (Smith et al, 2017, cited in Yar and Steinmetz, 2019). Women may, in fact, be more likely to engage in online stalking behaviour than men, even while men are much more likely to engage in the blending of cyber stalking with real-world intrusions (Strawhun et al, 2013). Women are, however, significantly more likely to experience stalking behaviour (whether online or not) as threatening (Owens, 2016). When online intrusion is conducted as an extension of real-world stalking (a situation that is not always easy to accurately identify, given the potential anonymity involved in either or both areas), it is more likely to be experienced as threatening. Acquardro and Begotti (2019) found that the health and anxiety effects of cyber stalking were far higher when the victim had also experienced real-world stalking in the past, whether or not this subsequent cyber stalking was carried out by their former real-world stalker. Ahlgrim and Terrance (2018) found that identical scenarios were judged more threatening when the perpetrator was presented as male, while victims were considered less sympathetically if they were male. These results were significantly compounded by the gender of the participants being asked to evaluate the scenarios being presented to them (women finding male perpetrators more threatening than male participants). Finnegan and Fritz (2012) found very similar results, with respondents much more likely to identify the same scenario as 'stalking' if the agent was male and the focus of their intrusion was female.

The concept of stalking came to prominence in relation to celebrity stalkers in the 1980s, and, in particular, after an obsessive fan shot and killed the former Beatle John Lennon in 1980. Chris Rojek's (2016) *Presumed*

Intimacy explores the history of the phenomenon of celebrity obsession, and the migration of the phenomenon into an age of new media and of the supposed democratization of celebrity. Maggie Wykes (2007) suggests that the murder of the television presenter Jill Dando in the UK in 1999 was the first high-profile case of an obsessive fan using the internet to stalk a celebrity. It was suggested at the time that the killer had been able to identify the presenter's home address from materials posted online. Today, most people have some kind of social media profile and are, therefore, potentially subject to levels of stranger engagement that would have been impossible in earlier times. After John Lennon's murder, a number of other high-profile celebrity-stalking cases gained significant attention from both traditional media and politicians. This focus on celebrities did, however, lead to a greater attention being paid to the, until then, relatively underreported issue of the stalking of non-celebrities by ex-intimates, as part of a wider feminist critique of gender relations in society. Hegemonic conceptions of masculinity (Connell and Messerschmidt, 2005) that celebrated and encouraged, as romantic, male sexual conquest and pursuit of women in the face of resistance to such advances (that is, not taking no for an answer as 'faint heart never won fair lady') came to be seen as unacceptable in the 1990s. As such, the rise of the internet coincided with what Anthony Giddens (1992) called 'the transformation of intimacy', from relations based on traditional gender roles and expectations about agency and passivity, towards plasticity, negotiation and consent. In the early 2000s, it appeared that the affordances of the internet, in terms of concealment, evasion, remote access and the ability to incite at a distance, would mean persons who would not engage in stalking in real-world situations and fashions might be willing to do so online, as inhibitions to do so were minimal (Bocij and MacFarlane, 2003).

It is paradoxical that the rise of new media coincided with a cultural shift towards challenging traditional gender roles and hierarchy. This shift lead to an unprecedented emphasis upon informed and explicit consent in the formation and ongoing management of all relationships. Yet at the same time, social media facilitated the rise of a form of micro-celebrity culture, in which greater and greater numbers of people are now presenting themselves online to others in a way that is then increasingly hard to control in terms of subsequent access to and interpretation of this material. The use of the term 'stalking' to refer to looking at someone else's Facebook page ('Oh, I've been stalking you on Facebook recently') should not confuse us. To look at a public social media profile does not qualify as 'stalking', as it is neither unwanted nor intrusive as such. However, placing personal materials in the public domain can afford unwanted intrusions. When ease of access increases, so does victimization (Navaro and Jasinksi, 2012). While Pittaro (2007) suggests online stalkers with a degree of computing skill can easily evade detection by victims, and while the scale of individuals holding the skills necessary to do so has increased over time, this has occurred even as the ability of authorities and service providers to identify such stalkers – reducing the relative ability to evade and conceal if one is not increasingly competent. Increased exposure (posting personal material) and proximity (allowing open access) online does increase the risk of being cyber stalked; but evidencing links to guardians (digital and in person) reduces targeting (Reyns, Henson and Fisher, 2011). The same risk-taking factors that predict victimization also predict likelihood of perpetrating stalking behaviours (Wick et al, 2017). While tolerance towards offensive speech and communications has reduced, what is considered acceptable and unacceptable are not fully agreed upon in newly emerging spaces of online communication (Short et al, 2015), leading to conflict.

Marganski and Melander (2015) seek to extend research beyond cyber stalking to a broader conception of cyber aggression and what they call 'online obsessive relational intrusion'. They found most female student respondents in their survey had experienced one or more of the activities they asked about in this regard. However, where almost half of respondents answered yes to questions concerning partners purposefully ignoring them online, repeatedly asking where they are by electronic device, saying something to deliberately anger them or intentionally making them feel bad by electronic message, only between 4.1 and 6.5 per cent of respondents said they had experienced threatening communications from a partner, or had private information or images released online. Further, 42.6 per cent of respondents said a partner had looked through their phone without permission. While such actions might associate with intimate partner violence and stalking of ex-intimates, on their own, they do not predict it in the majority of cases.

The concept of bullying has also undergone a significant transformation over time. Once a term reserved for harassment of children by other children, it has expanded, in respect of both target and perpetrator. Today, it is not only children at school or at home that are seen as potential victims of bullying (from other children). Adults are also seen to be potential victims of bullying in the workplace, or elsewhere, if they are subject to forms of humiliation, violence and/or the threat of violence by other adults, whether or not those others are in positions of authority or are peers. Today, it is said that parents may be victims of bullying by their own children. Levels and forms of authority and even physical punishment that were once considered legitimate, necessary or even virtuous are now seen as unjustifiable forms of bullying. Where the internet promotes horizontal interactions and challenges notions of hierarchical authority, it has encouraged the view that no one is better than anyone else is, nor has the right to constrain or punish anyone else. At the same time, new media promotes horizontal forms of 'unauthorized' criticism that are now labelled as bullying. As such, the internet radically reduces tolerance for something that it also affords the incidence of, and incites at a significantly increasing rate.

Whether the internet increases or decreases the capacity to evade and conceal those engaged in stalking and bullying is disputed (see next section). What is certainly true is that the digital has afforded new means of accessing people. On a par with stalking and bullying, trolling is a form of group access, where a person engages in unwelcome and unwarranted access to a group forum with the intent to disrupt, harass or humiliate the members of the group they seek to communicate with online. While 'trolling' in this sense is a 'pure cyber crime', in Wall's sense of the term (in that it only exists online), other forms of collective disruption and harassment, of course, exist in the physical domain. Whether trolls are intrinsically more disruptive, due to greater ability to access, conceal, evade and incite online, is another matter. While harm certainly can be done, online forums that are disrupted by trolls can and have fought back, as have the victims of cyber bullying and stalking.

Seeking to present oneself online as someone else is a form of fraud and identity theft. When conducted for the purposes of stalking, trolling or bullying, such identity theft is referred to as 'catfishing'. Such behaviour seeks to utilize all key affordances of the digital domain, to access, conceal, evade and incite. While identity theft has always existed, the separation of identity from the identification that is key to all forms of online interaction does make identity theft easier; but that is not to say it is easy to maintain indefinitely, however disruptive it might be to be the target of such behaviour.

2. Is the disappearance of disappearance 'symmetrical'?

David Wall (2007, p 36) refers to digital 'mouse droppings'. While using the internet may feel anonymous, as users are interacting with machines in the absence of any other persons in their physical environment, the searches people carry out and the messages sent/actions carried out are always in some sense recorded – and, in almost all cases, such recording is retained for a significant amount of time. Wall refers also to Haggerty and Ericson's (2000) notion of the 'disappearance of disappearance' in this context. What may seem like a fleeting action and the click of a button, with no lasting consequence, is in fact recorded. Where real-world interactions, whether verbal or physical, are rarely recorded and so, in many cases, leave no 'proof' that they ever took place, actions and interactions online almost always leave a trace – what Wall calls digital mouse droppings. Tracing the trail of such mouse droppings makes it possible to detect criminal acts in a way that was far harder in the world of physical forensics.

In the past, telephone conversations were rarely recorded unless a prior decision had been made to 'tap' a particular line. Today, digital telephone calls are always recorded, as are the 'meta–data' of who called whom and when (again, something that was rare in pre-digital times). Disputes over when such evidence can be used in criminal cases now takes place, as do disputes over how such data should be retrieved and from whom. In January 2010, after Robin Hood Airport near Doncaster (England) was closed by snow, a prospective passenger sent an angry tweet, claiming he would blow up the airport if it was still out of action when he was due to fly (some ten days later). In May 2010, the man was convicted of sending a 'menacing electronic communication' and fined. The conviction was later overturned, on the basis that the tweet could not have reasonably been taken as a real threat (though the original decision had seen it as such and was only taken to appeal when a number of celebrities paid tens of thousands of pounds for the accused's defence costs), at least according to the appeal judge. If the police and prosecution services had not investigated and brought a case, and if the man had, in fact, subsequently carried out his threat, the existence of the prior tweet would have been taken to show that the authorities had not acted upon evidence, and so could have been blamed for any subsequent harm. As such, authorities are under pressure to investigate and act upon every recorded interaction that might imply a threat of some kind. Given that almost all digital interactions are recorded (and so count as 'evidence'), this creates an escalating scope, not of errors as such but of recorded evidence of that error occurring.

When authorities are found not to have acted on evidence, or where evidence would have been available if sufficient attention had been given to what data had been collected, those authorities are said to have failed in their duty to protect citizens. The possibility that sufficient combing through various records might have highlighted a pattern of unusual behaviour by those who later carried out the 11 September 2001 (9/11) terrorist attacks in the United States is a case in point. However, after 9/11, the huge increase in resources dedicated to looking for such patterns has itself been questioned on the basis that these additional powers represent an undue intrusion into the privacy of citizens. Regarding the 'disappearance of disappearance' in relation to stalking and other forms of interpersonal harassment and abuse online, this tension has become increasingly evident in recent years.

Emma Jane (2016) discusses the case of Alanah Pearce, an Australian computer games journalist, who, in 2014, received sexually abusive threats via Facebook. Pearce's response was to trace the identities of the perpetrators, who turned out to be teenage boys. She then contacted the mothers of the boys concerned, asking whether they knew what their sons had been doing. Pearce went on to reveal some of these communications – revealing how mothers had responded and had set out to discipline their children. Jane outlines the way Pearce's actions were celebrated as a form of digitally correct 'digilantism' (or 'digital vigilantism') by mainstream media and politicians alike. While partially supportive of Pearce's strategy, Jane questions the extent to which such a strategy of victim reaction is either sufficient or appropriate, asking why it should be victims who have to tackle perpetrators, and whether it is always as easy as was the case with teenage boys using a public platform like Facebook. Older trolls and bullies have access to more sophisticated systems, to render their communications less easy to trace. Jane also questions the idea that mothers are best targeted as 'responsible' for their sons' actions. Nevertheless, Pearce's actions clearly indicate the capacity to respond to online abusers. If she could so readily identify her abusers, then technology firms cannot hide behind the claim that identifying perpetrators is somehow impossible. It is not. The extent to which some platform users' privacy should be protected relative to the safety of other users is a matter of policy decision, not something dependent upon any technical limitations of the platforms they operate.

Corinne Mason and Shoshona Magnet (2012) outline the paradoxes of digital technologies in relation to stalking and other forms of abuse against women, such technologies being both an instrument that can be used to track victims and a means of tracking perpetrators as well. An integral part of many social media platforms and other digital service providers (such as Apple, Facebook/Meta, Lexus/Nexus and Thomson Reuters) is the embedding of tracking cookies within the apps that many users take for granted. In the earlier years of the 21st century, digital technologies like a GPS device might be attached to a victim's car to enable a stalker to track a person's movements, or a mobile phone might be left on silent mode in such a way as to act as a recording device. More recently, all these features can be operated via the victim's own phone without their knowledge. A victim's phone can act as a tracking device, and various apps can be used to locate a person in real time. Mason and Magnet also note that work with companies like Apple and Google has brought forward some positive developments, such as removing the location of anonymous women's shelters from Google Maps, and creating apps that make it easier to tell trusted friends about one's location and safety situation. However, these authors also note that law enforcement in relation to violence has often been disproportionately targeted towards poorer and minority men, and that what they refer to as 'the prison industrial complex' cannot always be trusted to decide who needs to be kept safe from whom. While Jane (2016) argues victims should have the right to be protected by society, Mason and Magnet (2012), though agreeing, also point out that it is sometimes 'society' that victims need to protect themselves from in the first place. The question of how to deal with stalking behaviour involves multiple levels of action (Miller, 2012). Identification of successful strategies at one level (such as personal actions aimed at avoiding, detaching from or confronting perpetrators, or seeking assistance from guardians – personal, professional and/or legal) should not discount the need to combine strategies to avoid making the victim either responsible for dealing with their victimization or powerless in the face of it. Digital tools are part of the solution, even if they are also part of the problem. Nobles et al (2012) found that self-protection strategies were more often successful in cases of cyber stalking, relative to physical-world stalking, where victims were more likely to recourse to authorities. That it is not always possible to know whether cyber stalking is an extension of real-world intrusion means this distinction is not clear-cut.

The question of how seriously stalking is considered by police officers is significant. Lynch and Logan (2015) found that around 60 per cent of officers interviewed in their sample had never charged someone for a stalking offence, and those officers who had not done so were more likely to rate stalking as a less serious kind of offence than officers who had brought such charges. This reluctance, combined with lack of experience, is compounded in relation to online stalking behaviour, as it is even less likely to be within officers' routine police experience. When members of the public begin to get involved in 'policing' other citizens by means of digilantism, things become even more complex and confused (Trottier, 2012; Williams et al, 2013). At what point does naming and shaming online those deemed to be guilty of anti-social behaviour (whether online or in the real world) become itself a form of stalking and/or bullying? Sometimes, the best level of action for policing bullying and trolling online is that of the communities of users themselves, rather than external (real-world) policing agencies. However, these communities may sometimes have values that conflict with wider social norms. They may themselves come to exclude or marginalize those who do not fit in with their self-defined and valued idea of 'community' (Wall and Williams, 2007). Digital networks afford new forms of access and concealment, but the disappearance of disappearance (digital mouse droppings) can enable authorities to identify stalkers just as stalkers can access victims online.

3. Stalking online: changes and continuities, gender, youth and risk profiles

The idea of 'intrusion' is linked to the concept of 'privacy', and both have changed significantly over time, in scope and meaning. This will be discussed in greater detail in Chapter 6. For now, it is worth noting that ideas of reasonable privacy have expanded in recent years, in bodily, spatial, informational and communicational terms, such that today most people expect a level of privacy in all these domains that would not have been assumed in previous generations. That such expectations exist online is in tension with the increased capacity for intrusion by just such online channels.

The rise of what Castells et al (2007) refer to as mobile network youth, since the 1990s in early adopting parts of the world (Scandinavia, followed by the rest of Western Europe, Japan and the United States and, subsequently, other parts of Europe, the Americas and Asia, and now in Africa also), has very significant consequences for the relationship between freedom and security in a post-patriarchal family and then the wider world. The computer-networked mobile phone first began to diffuse to children and teenagers in Scandinavia in the 1990s, as a means of maintaining links between children and parents in conditions where traditional family formations were breaking down, and where both parents worked and did not always cohabit. A device initially given to children to keep them connected to parents also afforded those children a capacity to network among classmates, and then with strangers. Dependence on devices for connectivity to guardians then also became essential for maintaining increasingly digital friendship networks, but also increased scope for hostile digital intrusions. For a generation grown dependent upon such networks, such intrusions can be more harmful than might be the case for older generations for whom such networks are less central to their social lives. The generation of mobile network youth that were children in the 1990s are today entering middle age. Digital dependence among such 'digital natives' has been tempered by increasing digital awareness and competence. Greater reliance upon networks is balanced by a greater capacity to find digital solutions to challenges, and an ability to 'find one's tribe' online even where other forums and spaces are hostile. Williams (2007) suggests forms of online self-regulation are emerging, by which communities of self-identification are increasingly developing modes of self-protection and self-policing that offer some degree of protection from trolls and bullies online that may not exist in real-world contexts. Finding one's tribe online offers a degree of strength in numbers that many find unavailable in real life, where they may feel isolated from others with whom they can identify (and who will identify with them in turn).

The sense in which digital channels have changed the character of stalking, and the way they have also reinforced existing patterns of intrusion, can be seen in the research of Berry and Bainbridge (2017) and of Southworth et al (2007). Berry and Bainbridge document a wide range of prior research, looking at the characteristics of cyber stalking relative to real-world stalking behaviour. This array of prior research produces highly contradictory results. Some studies suggest the pattern of stalking behaviour online resembles that of offline stalking (mainly male ex-intimates stalking female ex-intimates), while other studies find this no longer to be the case. These differences tend to reflect sampling methods. Where random samples are used, similarity tends to be highlighted. Change tends to be shown where self-selected samples are involved (where those identifying as having been victims/survivors tend to be more likely to participate). In Berry and Bainbridge's own study, two-thirds of their sample were women. As a self-selecting opportunity sample, it would appear that women, who experience stalking as more threatening, are more likely to participate in research on the topic. However, what the research also found was that men in the sample were just as likely to have been stalked, even if their underrepresentation in the study may suggest men are, on average, less threatened by online intrusive behaviour – which, the study found, was not very often linked to real-world abuse. Online stalking, in most cases, stays online.

Where Berry and Bainbridge identify a degree of online stalking that is distinct from real-world stalking behaviour, Southworth et al (2007) focus attention on those cases where online intrusion is an extension of real-world stalking behaviour. Southworth et al are keen to avoid the very term cyber stalking for the reason that it is both too narrow (in focusing only on computers, where multiple other digital devices might be used) and because it is too broad (encompassing forms of purely online interaction, rather than focusing on the way technology can be used to further pursue real-world abuse). Southworth et al prefer to use the term 'stalking with technology'. Whether 'purely' online intrusions should be labelled 'stalking' alongside forms of intrusion that involve physical forms of threat and intrusion is not simply resolved. Just because most online intrusion does not extend from or subsequently extend into attempts at physical contact, this should not make us forget that sometimes it does – and the threat of so doing can create fear. Also, if online intrusions act to limit a person's actions online, this is still a significant harm, especially in a society where online interactions are increasingly central to everyday life. Waasdorp and Bradshaw (2015) found a quarter of adolescents they surveyed had experienced bullying, while less than 5 per cent believed they had been cyber bullied. If they had, most believed their cyber bully was someone close to them in real life, even if they could not always be sure of this.

Bonanno and Hymel (2013) sought to identify the relationship and impact of cyber bullying to and upon bullying in real-world situations. There is, indeed, such an association, as those who engage in one form are at an increased likelihood of involvement in the other. It is also the case that predictors of one are also predictors of the other. The effects of each are similar. However, the two are not simply one. There is, nonetheless, a greater likelihood of engagement in cyber bullying and of being a victim if either party is involved in real-world forms. As such, cyber bullying is 'additive', in both the scale of events and the consequences of them. Just as was found by Berry and Bainbridge (2017), regarding the variability of estimates concerning the extent of cyber stalking (due to variation in sampling and data collection methods), so Brochado, Soares and Fraga (2016) found large variation in estimates of cyber bullying. Their 'scoping review' looked at 159 empirical studies seeking to measure the extent of bullying online. Significant variation existed between countries, with North America (the US and Canada) scoring highest overall, but there was also very significant variation between studies within the same country. The authors suggest this may reflect public awareness varying between years within countries and altering self-perception regarding bullying, but also data collection methods and sampling. Differences in the duration of recall periods (asking whether you have experienced bullying online in the last six months or one year) may explain some variation. However, variation within countries controlling for recall period remained very high (variation between 1–3 per cent and 50–70 per cent being common). Nonetheless, even if definitions are fluid, experiences remain real and significant. Pabian and Vandebosch (2016) found that social anxiety is more likely to drive forms of online bullying than is the case in real-world victimization. Those with higher levels of social anxiety are also more dependent on online communications at the same time as being more prone to victimization, creating a vicious circle of escalating anxiety, hostility and social dissociation.

4. From the physical to the digital, escape or extension?

Rice et al (2015) studied over a thousand middle-school pupils (aged between 12 and 15 years old) in Los Angeles schools, to look for predictors of cyber bullying perpetrating and victimization. Those that spend more than three hours a day online were at highest risk, alongside those who sent more than 50 texts a day. Girls were more likely to be both perpetrators and victims, while sexual minority pupils were at significantly greater risk of being victims.

Mishna et al (2012) carried out a questionnaire survey of children in middle and high schools within a large Canadian city, generating a sample of 2,186 cases. Their focus was on the relationship between age and gender, as well as the likelihood of being a victim or perpetrator of cyber bullying or of being both victim and perpetrator within the three months preceding the data collection. Results suggested that while 30 per cent of the sample had been either a victim or a perpetrator in that prior three-month period, around 26 per cent had been both – more than had been either one or the other. Female students were more likely to be in this 'both' category, which contrasts with real-world bullying, where girls tend to be either one or the other. Being party to online bullying also correlated strongly with real-world bullying. It was also the case that lower-aged students (aged 12–15) in the sample were less likely to be party to cyber bullying than were high-school pupils (15–18). Older pupils were better able to engage with technology, and so, it seemed, more likely to use it to bully peers. Parental engagement with their children's computer use also correlated negatively with the likelihood of being involved in bullying; and, as parents were more likely to monitor younger children's computer use, this may explain reduced levels of bullying. It was also the case that giving friends account passwords was a strong predictor of being bullied. It appears that these pupils (aged between 15 and 18 years old) fell between reduced parental guardianship and a limited development of self-guardianship, in terms of giving away passwords to peers. Hempill, Tollit and Kotevski (2014) also found that girls were more likely to be perpetrators and victims of cyber bullying, even as boys were more likely to engage in real-world bullying. This study, from Melbourne in Australia, found that prior experience of real-world bullying was strongly associated with cyber bullying in later school years, and that a disrupted home environment was the strongest predictor of real-world bullying. The researchers also highlight that prior research has indicated that school exclusion or a poor relationship with education also predicts real-world bullying and victimization, which then predicts later cyber bullying. As in Mishna et al's (2012) findings, lack of strong relations with guardians is key to young peoples' vulnerability. Kowalski and Lumber (2013) found that among school pupils aged between 11 and 18 the impact of cyber bullying was significant, not only in terms of mental health but also on physical health and academic performance. Interestingly, these authors discovered that the group that suffered the worst impact on all three counts were those who were identified as being both perpetrators and victims of cyber bullying (see also Kowalski, Lumber and Agatston, 2012).

Whereas Castells et al (2007) suggest that a generation of 'digital natives' do mature over time to be better able to navigate the online world, relative to their experiences as high-school pupils, Nycyk (2015) suggests that victimization is not so easily overcome with age and experience. Nycyk (2015, p 12) argues: 'The misconception is that adults are better able to manage it.' He cites a number of studies regarding the negative impact of cyber bullying on adults: psychological distress, lowered self-esteem, weakened ability to form and maintain friendships, lower marriage rates, reduced employment and earnings, reduced health, depression, leaving relationships and work, self-destructive behaviours, increased alcohol and drug use, self-harm and even suicide. While schools and other educational institutions, and some large employers, have begun to develop anti-bullying strategies, and a wide range of self-help and treatment books, courses and programmes exist, those most marginalized by abuse are often those least likely to seek help or speak out. Nycyk notes that the rise of new media (Web 2.0) has intensified the nature of online bullying – via blogs, chat lines, email, Facebook, Instagram, gaming sites, phone apps, texts, Tumblr, Twitter, virtual communities and YouTube (2015, pp 6–7). Whether the array of alternatives represents escape when being bullied on one channel or, rather, an ever-increasing minefield of potential harm differs from person to person. For some, profusion is a positive means of increasing security, while for others profusion represents an increasing risk of harm, if not by one route then another. Abuse can take many forms: threats and name calling;

altering photos; creating fake, unauthorized or hurtful profiles; disclosure of personal information; flaming, hacking and desecrating of memorial websites; impersonating someone; posting gossip and rumours; and sexting (Nycyk, 2015, pp 7–12). Factors increasing likelihood of bullying online, Nycyk suggests (following Suler's disinhibition categories), are: dissociative anonymity, invisibility, asynchronicity, solipsistic introjection, dissociative imagination and minimizing authority. Strategies designed to reduce this sense of concealment and evasion will, therefore, reduce such behaviour.

Whether companies like Facebook (with its subsidiary companies, WhatsApp and Instagram) and Google (with its subsidiary YouTube), and Twitter do or do not do enough to limit the anonymity of bullies on their platforms is much disputed. These tech giants claim to be doing significant work to limit the scope for bullies, but critics challenge such claims. Danny Fortson (2021, p 31) points out that platforms like Twitter, Google and Facebook, along with their subsidiary services, actively target people with content that may be harmful even without intent. Whether this constitutes a form of automated bullying is open to discussion. However, if such platforms have automated a form of targeting that is strongly associated with reducing the self-esteem of users (particularly younger female users), it is hardly likely that such services are going to be very successful at removing bullying from their services. Such services might be said to have created a form of auto-bullying rather than their algorithms being used to eliminate it. Even if such platforms do invest in reducing levels of concealment, their own algorithms may themselves have become the greater invisible threat when it comes to auto-bullying.

5. Trolling of feminists

The trolling of feminist activists (such as in Gamergate; see Nagle, 2017) illustrates the potential for connectedness between interpersonal bullying and group-directed (and, hence, political) hate speech (Vera-Gray, 2017), but there is always scope for resistance. Fiona Vera-Gray recounts:

> [O]n a Saturday morning in June 2012, I started receiving notifications of abusive messages being left on the [a recently commenced] research site. Site statistics revealed the site was viewed 948 times on this Saturday. The comments continued throughout the weekend, culminating in a total of sixty-one comments left on the site by eighteen individual commentators. Thirty-six of these met the criteria for trolling (from fourteen commentators). (2017, p 69)

The internet has created a range of new spaces for feminist discussion and organization. In this sense, the internet has been positive for feminist activism (see Lewis et al, 2015 for a discussion of 'safe spaces', in terms of being 'safe from' as a precondition for being 'safe to'). However, such sites and spaces online have also become targets for trolls (p 65). The purpose of the research project Vera-Gray was seeking to undertake, and which was targeted, was a survey of women's experience of 'street harassment' or 'male intrusion' into women's lives in public spaces. The design of the research involved a preliminary conversation with participants, which sought to frame women's experiences in public spaces as 'space invaders' (in terms of proximity), 'the gaze' (where men looking at women is seen as intrusive) and 'verbal ejaculations' (uninvited talk and noises). Such framing preceded an invitation to keep a diary of such events, and a subsequent follow-up conversation. The pre-emptive framing (in terms of 'space invaders', 'the gaze' and 'verbal ejaculations') was aimed at 'subverting dominant narratives' (p 68).

Vera-Gray notes earlier research by Herring (2003, cited in Vera-Gray, 2017, p 66) showing that, while the most highly cited concern expressed by men regarding 'threats' relating to the internet was over loss of free speech, the most commonly expressed concern for women was in relation to loss of privacy. These contrasting concerns mean that some men feel threatened over their right to express their opinions (in 'subverting [the] dominant narratives', as they imagine them to be, of feminist criticisms of men), while some women feel threatened by these expressions, especially when such expressions are posted on platforms largely or exclusively designed for discussions among women.

Vera-Gray notes (pp 70–73) two types of 'trolling' in relation to the research project in which she was involved. One type involved overt threats and sexist insults. The title of her article is one such quote: 'talk about a c*nt with too much idle time' (note: Vera-Gray chose to keep the swear words intact); 'B*tch, please find the tallest skyscraper near you, and jump off it'; 'sounds like a snobby little princess … type, there's only one real way to cure them but its illegal' (sic). The second type of trolling was less overtly threatening, and sought to question the research methods, claiming that the research design was flawed and would tend towards 'confirmation bias', that it conflated multiple categories of 'attention' under the umbrella of 'abuse' (from compliments to insults and threats), and that it would tend towards sampling bias in the way participants were recruited. While the second type of comment was less overtly threatening, Vera-Gray notes that its association (and co-location) with the more violent and misogynistic comments makes it impossible to separate them out, and so all combine to try to marginalize feminist researchers online. In her subsequent book, *The Right Amount of Panic* (2018), Vera-Gray discusses the wider logic of not knowing, such that it is not possible to identify 'levels' of threat in the lived experience of intrusion so as to be able to tease out 'serious' and 'less serious' forms – whether these be online troll intrusions or real-world 'street-level' intrusions.

An important observation in Vera-Gray's (2017) article is that social media platforms have policies regarding trolling behaviour, but these can be readily circumvented. When content is posted on a social media platform, it must usually be vetted by the platform moderators or by the platform provider; but comments made about such initial content does not receive the same level of scrutiny, such that large amounts of offensive 'trolling' material gets through, despite the measures put in place to limit it. Lewis et al (2016) found that only 3 per cent of feminist recipients of online abuse were satisfied with the response of internet service providers, and only 10 per cent were satisfied with Facebook if they reported trolling to that platform. Most victims sought their own solutions, whether than meant confrontation or adopting security measures to limit access. Logan and Walker (2017) argue a more detailed approach to mapping risk is required to guide safety planning, both at the level of the individual victim and in terms of other responsible guardians who currently adopt a presumption of low risk unless compelled to think otherwise. Phipps and Young (2015) investigated how female students at UK universities engage with 'lad culture' online and face to face. While sexist behaviour and communications are problematic, what Phipps and Young point out is the agency of female students in standing up to it. While institutional responses may be weak and should be improved, it should also be noted that women are active agents and often successful in challenging sexism for themselves.

Angela Nagle (2017) offers up a dystopian account of the current state of online communications in terms of transgressive polarization. If, in the 1960s, the politics of 'transgression' appeared to be a progressive politics of personal freedom and societal liberation, today 'transgression' is as much about giving offence as it is about personal autonomy. Trolling on 4Chan and 8Chan internet channels has created powerful 'meme factories' for the production of new and increasingly offensive modes of communication, often in the name of 'freedom of expression'. As channels for the expression of hateful opinions about individuals and groups, such channels have created levels of linguistic hostility that are designed to create a reaction and to, thereby, polarize discussions and celebrate the ability to offend or 'cancel' others.

Nagle writes:

> One can feel the life draining out of the body at the thought of retelling or rereading the story of the gamergate controversy, one which involved internal controversies, hit pieces, hate campaigns, splits and a level of sustained emotion more fitting to the response to a genocide than a spat over videogames. (2017, p 19)

Gamergate arose out of a dispute over *Depression Quest*, a computer game dealing with depression, developed by Zoe Quinn. Disputes arose over reviews of the game, some positive, some negative, with supporters and detractors claiming those with the opposite view to them were biased. The dispute escalated, with critics of the game becoming increasingly antagonistic towards Quinn, making and circulating hostile claims about her, and issuing threats. Literally millions of people became involved in attacking Quinn, based on numerous questionable claims made about her. In essence, Gamergate exposed a deep-seated hostility to women in the games industry and among many, many 'gamers', for whom Quinn's game about navigating mental illness came to symbolize the opposite of their own preferred gaming themes of violence, war and symbolic domination.

Anita Sarkeesian's feminist YouTube videos, setting out a feminist media critique of violent content dominant in video games, saw her being flooded with hundreds of thousands of hostile posts, and even a video game where the only object of the game was to punch a digital representation of Sarkeesian in the face. Sarkeesian's Wikipedia page was vandalized with pornographic content, and personal information was circulated, along with incitements to do her harm. Nagle goes on to document how alt.right activists have appropriated what was initially a form of geek/nerd transgression on 4Chan and 8Chan forums, into a political form of trolling – this, thus, linking back to what has been discussed in Chapter 2 on the topic of hate speech. Where left-wing campus politics in the early years of the 21st century became increasingly consumed with what Nagle calls the management of scarce virtue (increasingly strong claims to being offended and demands to cancel those who are seen to have created such offence), alt.right activists have mobilized resentment towards such 'cancel culture' in a politicization of anti-social and offensive 'lols' behaviour and chat online. What Nagle refers to as 'Gramscians of the alt.light' are cultural conservatives (most notable among them is Steve Bannon), who have actively sought to mobilize an identity politics designed to feed off left-wing offence, promoting themselves in such a way as to purposefully generate prurient reactions that can then be used to bolster conservative resentments in the so-called 'culture wars'. It has become increasingly difficult to 'respond' to such trolls, as a response is precisely what they are seeking to provoke, even as not responding offers little solution either. Trolling is a digitally distinct form of access aiming to incite conflict. Digital concealment and networked evasion afford such novel intrusion, but access can be symmetrical. Trolls are often easier to identify than their technological over-confidence might lead them to believe.

In conclusion

Stalking, as a legal term, is relatively modern (arising first in California in the 1990s, and only legally acknowledged in the UK in 2012), and it has both expanded its meaning, from real-world intrusion to the digital, and expanded in scale, with the escalation of social media dissemination of personal information online. Bullying has, likewise, extended into the digital. In both cases, online abuse is distinct from that in the real world in most cases, and does not routinely overlap, but when it does, this compounds harm. The disappearance of disappearance means the increasing difficulty for victims maintaining their own privacy, but, at the same time, perpetrators are less able to conceal themselves from guardians, if such guardians do set out to identify them. Finding one's tribe online creates new opportunities to escape real-world isolation and victimization. However, the polarization that is arguably implicit to the forming of such distinct online communities has increased the tendency for any interactions that do take place between communities to become increasingly hostile – such as in the form of trolling. Access and incitement polarize in their symmetrical adoption by competing parties, even as concealment and evasion are themselves contingent in their asymmetry.

PART II

Obscenity

4

Pornography and Violent
Video Games

Key questions

1. When, if ever, should lawful acts between consenting adults be unlawful to display?
2. Can different legal systems maintain their own definitions of obscenity in a global, networked world?
3. Can simulated acts of violence corrupt, or otherwise harm, those that view them?
4. How should a balance be struck between protecting people from their own preferences and the right to engage in and display actions that others consider wrong?
5. Does watching pornography and/or playing violent computer games incite or defuse aggression and/or frustration?

Links to affordances

Global digital networks certainly increase access to pornographic content, but that is not to say that certain material is not blocked, or at least blocked to a degree. Legal differences between states create scope for evasion, but laws have been significantly harmonized in recent years. Countries with the capacity to block and trace non-consensual content do so (posing a challenge to efforts at concealment), even if the resources here may not always match that invested in seeking and circulating it. In relation to both non-consensual (or simulated, non-consensual, 'extreme') pornography and violent video games, content does not simply incite viewers to violent thoughts, actions and beliefs – even if such content may reinforce and/ or normalize existing dispositions and ideas in certain individuals. Access is the primary affordance of interest here, while concealment and evasion are in a constant state of cat-and-mouse between guardians and those who breach obscenity laws. The question of incitement remains complex, contradictory and contested.

Synopsis

Digital networks increase scope to access explicit sexual and violent content. The question of concealment and evasion take on a distinct meaning in relation to non-consensually circulated ('revenge') pornography, where what was thought to be private is disclosed without consent. However, regulation is possible here precisely because those that disclose such content are often not as anonymous as they believe they are.

Where issues of access, concealment and evasion are very significant in relation to pornography online, the question of incitement remains key to discussions of both sexually explicit materials and violent online gaming. Does the consumption of such material incite real-world acts of violence, from sexual violence to mass shootings? The full history of the law regarding the concept of the obscene is beyond the scope of this work. These debates echo longstanding disputes over 'media effects' in relation to violent behaviour; and these disputes remain polarized because the same evidence can be read in very different ways – not just in terms of how associations relate to causation, but in terms of how potential for harm should be balanced against freedom of expression at the level of the law and technical regulation.

Such debate around incitement is complicated by divergent affordances, as networks enable new scope to explore and validate diverse sexual desires and fantasies. This may be liberating for some, but may promote hostility from or towards others.

Chapter sections

1. The history of pornography, technology and social concern. The idea of 'obscenity', that which the sight of would corrupt and deprave, has exercised societies in different ways, but the ability of technology to circulate explicit images beyond established boundaries and narratives has been the prime driver of concerns for centuries. The internet is only the latest iteration, but it may be the most extreme (Wall, 2007). Concern is expressed from conservative quarters that technology allows pleasure to be detached from moral responsibility. Advocates of pornography, meanwhile, see pleasure being detached from consequences in a more positive light.
2. Non-consensually circulated (so-called revenge) and 'extreme' pornography. Where sex has been substantially decoupled from reproduction as such, the issue of 'consent' has replaced moral codes focused upon regulating sex within the child-rearing institution of

marriage. Obscenity laws have been streamlined, leaving legal concern to focus on non-consensual sex. Many formerly illegal forms of non-reproductive sex have been decriminalized, while the category of rape (non-consensual sex) has been expanded. Within the field of pornography, the law now focuses upon non-consensual acts or their simulation, in terms of 'extreme' (Attwood and Smith, 2010) or 'rape porn', and non-consensual display, in the form of 'revenge porn' (Calvert, 2015; McGlynn et al, 2019). (Non-consensual capacity in relation to children is discussed in the following chapter.)

3. Media-effects literature in relation to adult pornography. Atkinson and Rodgers (2016) refer to the 'pleasure zone' and 'murder box' when they describe online pornography and violent video games. They argue such virtual scenes create zones of exception that suspend moral norms, which can then seep into the wider culture. Critics argue this account assumes negative effects that it cannot actually show exist, and that pornography may be a positive influence in sexual socialization, exploring sexual identity and overcoming prejudice (McKee, 2010).

4. Media effects in relation to violent video games. Responding to Atkinson and Rodgers, Denham and Spokes (2019) argue that violent video games can foster forms of pro-social action, even if this is not in all cases the outcome of such game play.

5. Can the internet be regulated anyway? It is a common view that digital networks make the circulation of pornographic content impossible to stop and that laws and society have changed accordingly (Presdee, 2000). Societies and legal systems can and do, however, make choices that can have regulative impact (Hornle, 2011; see also Vera-Gray et al, 2021).

1. The history of pornography, technology and social concern

The *Oxford English Dictionary* defines pornography as: '1. the explicit description or exhibition of sexual activity in literature, film, etc., intended to stimulate erotic rather than aesthetic or emotional feelings. 2. literature etc. characterized by this [from Greek *pornographos* "writing of harlots"]'. This definition raises multiple difficulties when it comes to identifying what might or might not be referred to as pornography. In the first instance, there is the distinction made between the 'erotic' on the one hand, and the 'aesthetic' and/or 'emotional' on the other. If, following Kant, we define the 'aesthetic' as 'sensuous' experiences that evoke contemplative appreciation, rather than 'sensual' desires to use or use up (consume or consummate), then the distinction between art and pornography is tautological, but it is also one that is neither logically necessary nor empirically factual. What does it mean to distinguish the sexual from the emotional, when desire is an emotion – unless, by emotion, we seek to valorize some emotions, such as compassion and empathy, over desire? This may be to introduce moral distinctions between good and bad emotions, but on what basis should sexual desire be deemed negative? Perhaps, if sexual desire is fused with violent and aggressive emotions, such depictions might be viewed as immoral, but does that then mean that other explicit sexual material should not be viewed negatively? Might 'explicit' refer us to simply the revelation of things that may be harmful, not because they are intrinsically bad, but rather harmful when put on display? This raises the question of 'obscenity' (see sections two and three that follow). Also, what the creator intended in their work, and what a person viewing it thinks or feels, may diverge, as might the interpretations of different viewers; thus, defining pornography in terms of 'intended' effect is problematic, not least when the significance of pornography in society expands the further it circulates beyond its point of production.

Having its origins in ancient Greek, as 'writing of harlots', also locates the term 'pornography' in a very specific historical context. Whether the separation of the aesthetic and the erotic has any universal meaning across cultures and across historical time is highly questionable, and to impose any such assumption would be simply to assume one culture's values are universal, and that 'one' culture had a singular interpretation of its own words over time (which is also false).

It can be argued that the history of human culture begins with symbolic depictions of the world, but as language pre-dates its written form, the oldest artefacts demonstrating such culture are cave paintings and carved/shaped artefacts – one of the oldest being the so-called Venus of Willendorf, a female figure created around 25,000 years ago (McDermott, 1996). If we view this figurine as a fertility symbol, does that make it 'erotic' and, if so, does that make it 'pornographic'? Does that simply project a particular contemporary interpretation, or does it add to our understanding of the object and its creators? Ancient Greek and Roman celebrations of physicality, including sexuality, produced an array of material designed to stimulate sexual desire (whether by means of writing, drawing or sculpting) but which was not deemed morally questionable. References to or assumptions of pornography as material 'written by harlots' more likely reflects misogyny than a moral objection to graphic sexual content as such. Late 20th-century filmic reconstructions of Medieval European and Islamic Golden Age literature, such as Boccaccio's early Italian *Decameron*, Chaucer's Middle English *Canterbury Tales* and the Arabic collection of *One Thousand and One Nights*, present these texts as evidence of a sexual freedom before the strictures of subsequent forms of puritanism. Others have, in contrast, taken each of these films to give an aesthetic licence to what was simply cinematic pornography. Whether modern societies 'repressed' sexual desires or fuelled them with strictures is an ongoing dispute (see Marcuse, 1955; Foucault, 1976).

The modern use of the term pornography emerges in the 18th century, to describe the array of sexually explicit sculptures, signage and literature discovered when archaeologists first began to uncover the material culture that had been buried under volcanic ash at Pompeii after the eruption of Mount Vesuvius on August 24 79 AD (Varone, 2001). Initially, the materials discovered were kept in secret museums, only accessible to those wealthy enough to go on the kind of 'grand tours' that rich (mostly male) members of the elite could afford to take in the completion of their education. While challenging to the Christian conceptions of sex and sexuality dominant at the time, such material was not considered a serious threat to social order, as it was contained within physical museums and framed within a sense of historical distance. This changed with the advent of photography, which allowed such material as was being called pornography to circulate both beyond the elites and beyond the frames placed on such material by aesthetic and moral authorities (Bakker and Taalas, 2007). It is with the advent of photography that pornography comes to be associated primarily with visual reproductions. This is in distinction to sexually explicit literature (which, until the 19th century, precluded the vast majority of people in their lack of literacy). Concern over photography was also in distinction from sexually explicit paintings and sculptures (which, again, were far beyond the reach of all but a tiny minority of the population – a minority that, as today, worried about the capacity of 'others' to cope with content they themselves believed would not harm them). Photographs could be mass produced and illicitly circulated, and they could be seen as more 'real', being more immediate depictions of real people's bodies. The photograph, as unmediated media in both what it depicts and how it is circulated, begins a cycle of technological revolution and moral reaction, which develops later in the contexts of cinematic film, sound and colour film, subsequent post-cinematic video (extending from the public cinema to the private home), and then digital production, storage and distribution – taking access at each stage to ever greater levels.

Just as pornography became a moral problem when photography removed content from the frames of galleries and museums, so the idea of geographical context produced new contradictions (between distance and immediacy). Angela Carter (1985) notes the Victorian European elite's ability to locate nudity in either the 'compensatory ideologies' of the innocent (fairies and nymphs) or that of the exotic (African sculptures, Indian, Chinese and Japanese literature and art, and in European artistic depictions of Tahitian natives) so as to 'appreciate' the sensual/sensuous in the aesthetic domain, even while promoting puritanical sexual values in 'everyday life'. However, such boundaries were already breaking down, not just through unmediated media (from photography onwards), but also in terms of the movement of people, things and beliefs, across national borders, and between public and private spaces. For example, the Obscene Publications Act 1959 in the United Kingdom (see Yar and Steinmetz, 2019) reiterated earlier UK legislation that defined and prohibited 'obscenity' in terms of things likely to corrupt and deprave in a context of increased physical and media mobility across borders. The very attempt to draw such distinctions drew attention to them, and led to publicity being given to such works as D.H. Lawrence's *Lady Chatterley's Lover*. Just as in the 18th century, when elites travelled to Italy to 'complete their education' (including visits to the secret museums of pornography discovered at Pompeii), so now it became common practice on visiting France to buy a copy of *Lady Chatterley's Lover* there and seek to bring it back to the UK. This led to multiple prosecutions for having the book when coming through customs. The lifting of 'the Chatterley ban' in 1963 reflected the tension between increased mobility (of people and cultural works) and attempts to regulate personal behaviour according to state-level moral codes. As John Tehranian (2016) argues, attempts to define obscenity in a workable legal fashion in the United States have been equally problematic in the 20th century, with divergent moralities and mobilities generating as much contradiction in efforts to define obscenity as are resolved by jurisprudential manoeuvring. Attempts to limit and prohibit often expose and incite interest as much as repressing it.

The notion of 'obscenity' (as distinct from pornography as such) refers to the idea that certain content will likely corrupt and/or deprave a viewer, whether or not the act depicted is itself 'immoral' in the sense of being harmful to the participants (or, somehow, to wider society). Whether certain representations are only 'obscene' when what is depicted lacks a moral or aesthetic context (a frame) is open to dispute. In making pornographic content, harm to the participant may arise from engagement in an act (such as when a person is coerced or injured in the conduct of making a recording) or from the non-consensual circulation of the recording of a consensual act (such as in 'revenge pornography'). (Both these themes are discussed shortly, in Section 2.) Obscenity, however, refers to the idea that harm may arise from the display of material that is not necessarily harmful to those depicted but rather to those viewing it. Of course, harm to participants and harm to viewers do not preclude one another, but neither scenario is necessarily true and neither necessarily corresponds to the other.

Whether 'context' reduces harm is another question that remains in dispute. Some critics of pornography argue that graphic sexual imagery is obscene because it moves sex out of the context of marriage, reproduction, love, modesty, fidelity and monogamy. More recently, prohibitions have narrowed around depictions of non-consensual acts (real or simulated) and non-consensual disclosure of explicit acts. As such, choice becomes the primary moral guideline. Some critics of pornography, however, argue that much harm draws from pornography remaining largely framed within cultural contexts such as patriarchy, heteronormativity and ideals of hegemonic masculine conquest and supposed feminine desire to be conquered. Advocates of pornography argue that it is precisely the ability to remove sexual pleasure from moral frames of reference that make it a positive and liberating phenomenon, even as it also removes the requirement to locate sex in the context of love, heterosexual reproduction and family formation. That some pornography frames sex in terms of hegemonic masculine scripts of power and conquest does not mean all pornography does this, and removing any single moral framing around sex does make it possible to explore alternative forms of sexual pleasure. Whether such scope to explore sexual pleasure outside of dominant moral frames tends to demoralize sex as such is open to dispute, as is the question of whether this is intrinsically a bad thing. Novels were once banned as they were considered immoral because they explored personal life in fantasy rather than simply expounding moral doctrine. Should novels be banned because they explore human imagination in the domain of fiction? Does fantasy that explores desire outside of moral bounds of necessity undermine morality, even encourage immorality? In his novel *Porno*, Irving Walsh (cited in Wall, 2007) argues that a consumer society demands the right to pleasure as the ultimate measure of moral achievement. The maximum happiness for the maximum number of people, a utilitarian concept of morality, rejects the notion that pleasure is the antithesis of morality; rather, it is its best measure. The development of the contraceptive pill gives greater control, in particular to women, over the relationship between sex and reproduction. Separating sex from reproduction also separates sex from institutional arrangements and gender relations associated with raising children (that is, heterosexual family formations). Pornography separates sex from reproduction in a different way. The moral consequences of this separation can be positive or negative. The downside of a utilitarian morality is that maximizing an individual's pleasure may lead to treating others as simply a means to an end rather than as ends in themselves. Walsh's suggestion is that a society where everyone chases after their own pleasure may be a society where, in the name of pleasure, people are all treated as objects. Digital mediation may also encourage people to represent themselves in objectifying ways in order to satisfy the demands/assumed desires of others (Yar, 2012). How should society and the law respond when people seek to objectify themselves?

2. Non-consensually circulated (so-called revenge) and 'extreme' pornography

The relationship between sex and technology is complex and always linked to questions of morality. Technology affords various kinds of connection and various kinds of decoupling. Such connection and decoupling afford new kinds of freedom and new kinds of control (both of individuals in relation to their own bodies but also in relation to how individuals relate to others and to society more widely). Digital technologies afford greater access to pornographic materials. Technology also enables new levels of evasion (in terms of assumed lack of consequence in relation to one's actions), and of concealment (insofar as individuals can, or believe they can, consume sexually explicit materials without being observed). Digital technology also affords greater levels of incitement. Content available today is far more explicit than that which was previously available and so is more arousing (even if there is much dispute over whether such intensified – *inciting* – material leads to real-world enactment of what is viewed or whether viewers are only incited in the realm of fantasy).

Just as there has always been debate over the consequences of contraception in relation to morality in the regulation of sex, so likewise there has been debate over the moral consequences of access, concealment, evasion and incitement to sexual experiences by digital means (rather than through physical interactions with people in the world). The contraceptive pill (and other forms of contraception), like pornography, decouples sex from reproduction. Some see such liberation as positive, while others see it as harmful as it detaches sex from traditional morals associated with reproduction, childcare and familial care. The development of contraception decoupled the relationship between sex and marriage, and made sex into something that could be engaged in 'recreationally' rather than something that would, most likely, bring obligations. Sex for pleasure outside of the institution of marriage has been viewed positively and negatively in just the same way as has been the consumption of pornography for pleasure. Contraception makes it increasingly possible to engage in non-monogamous sexual relationships with reduced risk of pregnancy – again, something that can be viewed differently depending upon one's moral perspective. The question of whether a moral code designed to regulate individuals in a world without contraception (and hence needing to bind individuals to the likely consequences of their sexual actions) remains valid in a world where consequences can be more readily limited is open to discussion. Pornography online creates a whole new level of complexity if access, concealment, evasion and incitement can be afforded virtually. Is consumption of pornography online 'only' masturbation, whereas actually having physical interactions with another person would constitute 'infidelity' if either party were in a relationship with someone else? If recordings of real people are being viewed, is that more 'real' than unmediated sexual fantasies? Finally, but no less significantly, decoupling sex and reproduction has significantly reduced stigma associated with non-reproductive sex in general and non-heterosexual sex in particular. When once masturbation and homosexuality were both deemed sinful because they separated sex from reproduction, today, sex for pleasure is more positively regarded (McCromack, 2012). If 'gay' sex means happy sex, then perhaps gay sex has become the ideal, rather than the pariah. Perhaps the struggle for gay liberation has liberated everyone, even if not everyone handles freedom very well. Experimental consumption of sexually explicit materials online has made explorations of difference something normal. Diversity is celebrated by some, even as it has challenged the moral frameworks that have previously sought to regulate sex.

The decoupling of sex and reproduction has gone hand in hand with a recoupling of sex with many assumptions about consent. It is worth noting that it has only been in recent decades that sex within marriage could be legally deemed as 'rape'. Until 1992 in the UK, it was assumed that the marriage contract, 'to have and to hold', constituted a lawful agreement to have sex, such that for as long as the contract was in force (until divorce, separation of annulment) each partner could reasonably expect sex, meaning that rape 'could not' happen. It was assumed that consent had been given at the point of marriage and that it was ongoing, unless undone by the undoing of the marriage and hence the contract. Legal adulthood assumes a person's competence to know their own mind when entering into contracts and that, as such, contracts are binding on parties. Until the ending of the contract, parties are assumed to be bound by their own decision to enter into that contract. Today, the concept of consent has been redrawn more narrowly, in terms of ongoing consent in time, and informed consent prior to entry into any relationship. Adults are assumed capable of giving consent in a far greater capacity than are children, and hence 'the age of consent' defines when a person can agree to have sex. In the UK, this was 13 until 1885, but was then raised to 16. The age of consent for gay men has twice been lowered in the decades following the decriminalization of male-to-male sexual activity, to now match that of heterosexuals. As with consent within marriage and in relation to children, so 'consent' has been narrowed in relation to what counts as 'informed' consent, such that intoxication, deception and non-violent forms of coercion, as well as actual or threats of violence, are now all deemed sufficient to make a sexual act non-consensual (see the Criminal Law Reform Now Network – www.clrnn.co.uk/ – Concent and Deception Project for more detail on what does and does not count as consent in the UK at the current time). How consent plays out in the online distribution of sexually explicit materials raises new and significant challenges to regulation, as repetition and distribution extend the significance of an act beyond any initial willingness to perform it.

The simulation of non-consent and the evasion of consent in online pornography take up the remainder of this section. The simulation of non-consent online refers to the production and distribution of violent sexual images. This can be said to be the display of unlawful acts even if the acts are agreed to or simulated (McGlynn and Rackley, 2009). Child pornography is the display of unlawful acts the production and distribution of which is a criminal offence. However, where images are of adults engaged in violent (or extreme) sexual acts, these acts may be simulated rather than real (in a way that acts involving a child, animal or a corpse cannot 'simulate' non-consent as such persons/animals cannot give consent to perform any sexual act). Just as violent acts are simulated in television and films by actors, and with the use of special effects, so it is that actors in pornographic content may simulate violence by pretending to perform violent acts or pretending not to consent to such simulated or real acts. However, is such 'pretence' real simply on the grounds of having consented to 'perform' it? Is 'violence' not real just because an actor agrees to it, or is it only 'fake' if it is entirely fabricated? Actors in violent films are not actually shot or stabbed, so is the sexual violence displayed in extreme pornography more 'real' when someone is slapped, choked or constrained? Is it more harmful because viewers might not realize that 'fabrication' is not 'documentary' (the supposed filming of 'real' events)? Attwood and Smith (2010, p 180) argue that believing that 'take[ing] pleasure in a "realistic" depiction of rape is ... incompatible with rejection of the crime of rape' is no more correct than believing that a woman who enjoys reading Nancy Friday's *My Secret Garden* (or, one might add, the even more popular *Fifty Shades of Grey* franchise) actually wishes to experience in real life what is being experienced in fantasy by reading such works. Attwood and Smith argue that 'legislation against the imagination' (2010, p 187) assumes the direct transmission of fantasy into imitation in reality, and is itself profoundly mistaken and dangerous (as it would licence prohibition of almost every creative work). On the other hand, McGlynn and Rackley (2009 and elsewhere) hold to the view that extreme pornography does 'cause' real-world violence, and that such content as does cause real-world violent imitation warrants the term 'extreme' porn (whether such circularity is taken as 'proof' or as 'tautology').

Where simulation of non-consent is one area of dispute, the converse is the evasion of consent such as when consensually produced, sexually explicit content is circulated without the consent of the person depicted. Brown (2014) notes that professional pornographers – for whom contracts exist – are increasingly challenged commercially by the rise of 'amateur pornography' and 'porn piracy'. Ironically perhaps, when set against non-consensually released 'revenge pornography' (or 'image-based abuse'; McGlynn et al, 2019), such professional pornography may come to be seen as almost virtuous, at least regarding formal and informed consent (in terms of explicit contractual terms and conditions). Revenge pornography, sexting and digital image-based abuse are a relatively new phenomenon that have challenged the law worldwide, with laws only being introduced during the 2010s, initially in the United States (California and New Jersey being first, with a number of US states still to introduce criminal or civil laws; Calvert, 2015) and then Europe, Australia, Japan and Canada, with some other countries subsequently following suit. The situation where a person shares an explicit image with someone they believe will retain their privacy, and that image is then more widely circulated, has been criminalized in some countries. Other countries have only criminalized the act of circulating an image without consent if the person circulating the image was the one who produced it. The first scenario involves the subject of the image initially sharing the content – and only some jurisdictions accept that this act of 'sharing' does not constitute an initial breach of privacy – so that further circulation by the recipient represents a crime. As social media platforms like Snapchat create the impression that an image circulated will be almost instantly deleted, sexting among teenagers has become normalized. Calvert cites one source suggesting Snapchat is 'like sexting with a virtual condom' (2015, p 699). However, such presumption of privacy is misguided, and the circulation of content can often breach the assumed intimacy of sender and recipient. Calvert refers to the suicide of Tyler Clementi in New Jersey in 2010, as the first high-profile casualty of such non-consensual distribution of sexually explicit material. McGlynn et al (2019) highlight a wider array of harms from such material (in terms of social rupture, constancy and isolation of victims). They also highlight ongoing problems with legislation over 'intent to cause harm', ineffective policing and initial disclosure by a victim being used to undermine their claim to having been victimized when what they initially shared is shared more widely without their consent.

While technology can give greater control over our own bodies, digital technology can also detach us from images of our own bodies (Petella-Rey, 2018). Pornography has decoupled sex and reproduction, and furthered a shift in moral discourse around sex, from a focus on regulating reproduction towards a focus on consent as the primary moral principle. As such, while access has increased, concern has shifted from access as such towards non-consensual access to explicit content, as well as the possibility that non-consensually released content or content simulating non-consensual acts might incite non-consensual sexual violence in those watching such material.

3. Media-effects literature in relation to adult pornography

The history of 'media-effects' research is usually presented as a series of extreme positions, followed by a 'now' where researchers claim to be more balanced. This 'pre-history' of reasonableness is usually said to have comprised three stages. First, the mass-society approach to newly emerging audio and visual media in the early decades of the 20th century suggested a 'hypodermic syringe' model of 'cause and effect' – such that the media injected people with beliefs and desires, which viewers and listeners were then said to accept and enact. The paradigmatic example of this (subsequently largely debunked) theory was the 1938 Orson Welles' radio production of H.G. Well's *War of the Worlds*, which newspapers at the time claimed caused mass panic and flight from US cities close to where the drama was set. There is little evidence that any such panic took place. Mid-century, 'two-step-flow' models (see Katz and Lazersfeld, 1955) looked at how media content is mediated by interaction with significant real-life others in the formation of opinions and behaviour. Later still, 'uses and gratification' approaches assumed that media consumers choose the content they read, view or listen to in accordance with their prior preferences, such that any association between media content and consumer attitudes and actions is caused by the latter determining choice to consume the former (and as such, no media effects at all). Contemporary new-media researchers claim to have moved on beyond such starkly distinctive positions, but such transcendence is in little evidence.

The question of online pornography's 'causal' impact on sexual behaviour, and, in particular, violent sexual behaviour, repeats longstanding disputes over 'media effects' in general. The rise of cinema was itself credited at the time with promoting anti-social behaviour, as had been the printing press in its day. The rise of the video recorder (and player) likewise saw concerns raised in the 1980s about 'video nasties' – often low-budget and censor-evading, home-video-market-oriented films, typically of a violent and/or sexual nature. Claims made at the time about the corrupting effects of such materials, particularly because they were distributed and watched in private homes (making them potentially available to children), were never proven empirically. The film *Natural Born Killers* (1994) has been said to have sparked multiple copy-cat killings by fans who have become obsessed with the film's anti-heroes. Watching the film multiple times has been associated with such subsequent killing sprees; but that a person would choose to watch the film repeatedly suggests something about that person/s, rather than suggesting that repeated viewing would have any particular effect on a viewer who did not have a prior violent disposition.

The murder of Jane Longhurst in 2003 highlights the same problem with evidence. Longhurst was murdered in Sussex, England, by Graham Coutts. Coutts had strangled Longhurst during a sexual encounter. Coutts' defence argued this had been the accidental consequence of engaging in what Coutts claimed was consensual (something the jury rejected) erotic asphyxiation. Coutts' computer contained many downloaded sexual videos of such acts. The prosecution maintained that these films contributed to Coutts' actions. Had such material 'triggered' his actions? Coutts' interest in erotic asphyxiation had begun long before he first accessed the internet, and his downloading of such material began many years prior to his killing of Jane Longhurst. Coutts had been treated by a psychiatrist for having murderous thoughts years prior to ever accessing the internet. Online material did not simply cause his actions, but perhaps such content as he actively sought out did reinforce existing fantasies, contributing to his enactment of them with fatal consequences.

Research into new-media-pornography effects is dogged by the same problem that has prevented any consensus on media effects generally. Laboratory-based controlled tests (such as the classic works of Albert Bandura and colleagues; see Bandura, 1963, for example) do often find causal links between exposure to violent actions/images and subsequent increases in aggressive feelings and actions, but are discounted by critics for being unrepresentative of how action develops in real-world contexts. Researchers who use alternative methods to research action in the real world (such as has been reviewed by McKee, 2010) tend to find no causal link between consumption of anti-social media content and harmful behaviours; but these studies are themselves discounted by those who believe in media effects, precisely on the basis that non–lab-based research findings do not arise out of controlled conditions.

Alan McKee (2010) outlines ten key points regarding research on the possible effects of exposure to sexually explicit scenes on children and on adults who were exposed when children. McKee argues this research is important in an age where concerns are increasingly expressed about the ease with which children can access pornography via the internet. The key points can be summarized as follows: curiosity about sex is natural and healthy; accidental exposure to real-world sexual behaviour does not harm children; age of first exposure to pornography does not correlate with subsequent negative attitudes towards women; and that current generations are no more sexualized than previous generations. McKee also carried out a retrospective questionnaire with over 1,000 respondents. Central to his findings were that the age at which a person was first exposed to pornography has fallen over the last six decades, with the digital being a very significant driver of this. However, attitudes to women among the male respondents showed consistent improvement over the generations, even as this correlated with earlier exposure to pornography. McKee is keen to stress that access to pornography cannot be said to have 'caused' improved attitudes towards women, but the evidence does discount the opposite view, which is that earlier exposure to pornography has any causal connection with negative attitudes towards women. McKee concludes that non-violent pornography can be positive in showing post-pubescent children that interest in sex and sexuality is normal and okay, even as accidental exposure to pornography by pre-pubescent children is largely meaningless to those viewing it. McCormack and Wignall (2017) go further and suggest that pornography is a positive and educational tool for many young men engaged in the exploration of non–heterosexual sexual preferences, and has a generally positive relationship with more accepting attitudes towards sexual difference.

Given the rather positive suggestions outlined, it is important to note that equally negative perspectives exist. Atkinson and Rodgers (2016) suggest that the digital domain creates 'cultural zones of exception' in relation to both sex and violence, the former being referred to by the authors as 'pleasure zones' and the latter as 'murder boxes'. (The latter concept is discussed in the next section.) Atkinson and Rodgers note that their research seeks to go beyond 'media-effects' research, in the manner discussed earlier. Given that media-effects research does not show any direct and simple causal relationship, to assert a negative consequence requires the suggestion of something more subtle and pervasive. Atkinson and Rodgers suggest that ongoing pornification of culture is creating a 'culture of spite' (Presdee, 2000) and a 'state of suspension', where empathy and compassion are replaced by a distancing of action and consequence, fuelled by online pornographic imagery. A sadistic voyeurism, these authors argue, is seeping into everyday life – what they suggest is 'the material consequence of fantasy' emerging from the mainstreaming of online symbolic violence and new social harms based on the cultural normalization of social subjugation and unfeeling destruction. Engaging in sexual fantasies based on detachment from care and consequence, and to select sexual experiences from online dropdown menus (through which a viewer can select, from among other things, the race, age, body features and sexual positions of their masturbatory object of attention) is, in the authors' view, inevitably going to coarsen real-world human interactions, sexual and otherwise, over time. The authors do suggest that commodification in general has created an emotional glaciation from which people seek refuge (labouring to feel) in intensified experiential spaces online, so it is not simply the case that digital sex 'causes' social detachment. However, as an escape route from the wider objectification that is capitalism, such online experiences as are provided by pornography can only (the authors argue) reinforce the problem. The kinds of sexual scripts, narratives and relations pervading even non-professional pornography, along with the framing of content and its menu selection, 'wallpaper our world' (p 1299) with 'the legitimation of predatory and abusive sexuality'. The authors abandon the language of 'effect' for what David Matza (1964) refers to as associational 'drift', that is, the gradual normalization of deviance – in this case, a view of sex as violent, emotionless, without compassion and simply a means of gratification without concern for others. That Atkinson and Rodgers' account does not match McKee's finding that levels of exposure to pornography correlates with positive attitudes towards women's autonomy and equality highlights that strong and contrary views remain. Nair (2010) notes that emerging digital affordances around pseudo porn, where digital editing can alter content to fake involvement in, or create non-real participants in, sexually explicit content, creates new and even more complex problems when it comes to defining harm, effects and how best to deal with them legally.

4. Media effects in relation to violent video games

In a review of research prior to 2010, Ferguson (2010) suggests that the negative impact of violent video games is more often published, cited by other academics and given media coverage; relative to research suggesting no effects or even pro-social effects. Ferguson notes that overall findings are very mixed, and that divergent measures of aggression, problems with third factors and deep confusion over associations relative to causes produced a field that was polarized and contradictory. Since that date, things have not improved, and I will here present a range of significant articles chronologically simply to avoid imposing a narrative that might imply any singular 'outcome' in what remains a highly disputed field. Sterner and Burkley (2012) identify the strong connection between graphic violence in video games and hypersexualized representations of women and offer a range of findings that suggest a link between playing such games and increased sexist attitudes and perceptions. Gabbiadini et al (2016) found that exposure to violent video games increased sexist attitudes and a loss of empathy towards female game victims among male players who identified with violent male game characters. Fox and Potocki (2016) found increased rape myth acceptance (RMA) among men who had higher levels of video game use, but that this relationship was itself mediated by levels of interpersonal aggression and hostile sexism. Higher levels of game play may be both the cause and effect of these 'third' variables, leading Fox and Potocki to suggest a 'cultivation' model rather than a direct causal relationship. Cunningham, Engelstätter and Ward (2016) question experimental and survey-based studies, noting that when new violent video games are released, general levels of violence temporarily fall. The association between violence and gameplay might then be one of catharsis, rather than cultivation. DeCamp and Ferguson (2017) found a weak but statistically significant association between playing violent video games and measures of hostility and aggression, but once predictors of prior aggression, and propensity to want to play violent games, were controlled for, this association disappeared (and in some instances reversed – again suggesting catharsis over cultivation). However, LaCroix et al (2018, p 413) later found 'increased hostile sexism' reported by players assigned randomly to playing a violent immersive video game where their opponent was a sexualized female character, relative to groups given a male opponent, a non-sexually dressed female opponent or after playing a non-violent game (Pacman). This result was conditioned by the participants' perceived sense of immersion, so may in part reflect the characteristics of players (the level to which they are taken in by the game) and the increasing capacity of games to immerse players. This suggests that for a certain kind of player, new levels of immersive gaming combined with hypersexualized violence may have a cultivating effect. A longitudinal study (Kühn et al, 2018) used magnetic resonance imaging and found that those exposed to violent video games were no different in their physiological (empathetic) reaction to another's suffering (when watching a film of someone cutting themselves while slicing a cucumber) than were members of a control group who had not played violent video games. The same team (2019) also found that with controlled groups, measures of aggression were no different between one group asked to play violent video games for two months, and a control group who were not. Kühn et al conclude that associations in other studies may be the result of 'priming', but if participants can be 'primed' by games researchers, might some not also be susceptible to 'cultivation' by games?

In Atkinson and Rodger's discussion of violent video games, they refer to the 'murder box' to parallel their earlier discussion of the 'pleasure zone' of online pornography. Both are said to be 'zones of cultural exception' but both are said to 'leak' into the world, or so the authors' argument seems to suggest even while it is repeatedly claimed this is not any kind of casual 'media- effects' theory. These authors continue their argument that zones of exception offer a vent for repressed desires that cannot be manifested within the constraints of modernity's civilizing process, while at the same time suggesting that violent video games are integral manifestations of modern corporate capitalism (the 'military-gaming complex', as they call it), both in games companies' profit motivations and in games' transformation of humans into objects to be exploited. The authors suggest such games as *Call of Duty (CoD)*, *Hitman*, *Battlefield*, *Assassins Creed* and *Grand Theft Auto* are based upon a logic of voyeuristic sadism, militaristic masculinity, the celebration of 'conquest and destruction' and 'destruction, murder, rampages, and bodily dissection or dismemberment' (2016, p 1301). Atkinson and Rodgers, while distancing themselves from laboratory-based causal models of effect, also reject what they see as the uncritical and celebratory research into 'gaming' that refers to itself as 'game studies' or 'ludology'. Whether violent game play vents frustrations generated in everyday life, vents frustrations that cannot be manifested (outside the box) in such a society, or fuels frustrations by generating aggression that might then manifest elsewhere, is never resolved; even as the suggestion of leakage is repeatedly alluded to. The authors argue the 'need for a scoping analysis of the way in which hegemonic norms and values are reproduced through the scripts and assumptions many games and the commercial logics that shape the nature of their content' create (2016, p 1302).

Denham and Spokes (2019) argue, in contrast to Atkinson and Rodgers, that, while not always the case, violent video games may encourage pro-social forms of action. Denham and Spokes carried out 15 'interactive-elicitation'-based observation/interviews with male and female players of *Grand Theft Auto*. The nature of the game's roles (and narrative structures/goals to be fulfilled) rewards violent actions and tends to minimize the consequences of gratuitous acts of violence; but players of the game also manifested resistances to these game logics, and often avoided violence in forms of pro-social activity within the limits of the game. Denham and Spokes do not seek to claim that such games as *Grand Theft Auto* promote primarily pro-social action, but rather: 'What we are arguing for is a more measured approach to player-game interaction when it comes to violence' (2019, p 751). They do, however, question Atkinson and Rodger's conception of 'zones of exception', noting that while such games offer elements of escape from 'dull consumerism', such games also contain materialistic goals that are often the narrative motivations for engaging in simulated violent acts (such as robbery). Atkinson and Rodgers argue (consistently or not) that, on the one hand, violent games both reflect and escape the instrumental rationality that pervades life outside of the box's 'zone of exception', even as (on the other hand) games may encourage the expression of irrational frustrations and aggression beyond the gaming space. Denham and Spokes suggest it is not the game that creates resistance to the constraints of everyday life. Rather, players of the game (at least sometimes) resist the constraints of the game's rules to bring pro-social motivations into otherwise anti-social gaming scenarios. Still, as Nagel (2017) points out in relation to 'Gamergate', when it came to trolling a games designer who resisted the logic of violence in video games, millions of gamers did step outside of 'the box', at least in terms of unleashing symbolic violence (and threats of real violence) against a real person on the other end of their digital messaging.

Wannes Ribbens and Steven Malliet (2015) add an interesting dimension to this discussion with their study of 26 players of violent video games. The researchers used observations of game play, diaries, focus group discussion and video commentary during play. The authors sought to study how players conceptualized and felt about the simulated deaths the game repeatedly confronted them with. Players adopted very different approaches to game play: focusing on narrative (following the story to transition through levels) or focusing on action (maximizing point scoring in simulated violent encounters); exploration versus mission completion; and reaction-based or strategic play. In essence, players adopted different rationales and emotional approaches. The authors suggest that certain attitudes and styles might increase detachment from feeling in relation to viewing mediated real-world violence, but there was no evidence that any style of game play would foster real-world enactment.

Matthew Spokes (2018) examined players of *Fallout 4*, a post-apocalypse survivor-based game in which players navigate within a world of death, corpses, skeletons and various dangerous creatures. Stokes refers to such games as 'critical dystopias' in that they encourage reflections on social cooperation and mortality. One player wrote: 'The other night I stumbled upon two skeletons huddled together at a broken bus stop. I imagined them as husband and wife, embracing each other for the last time as the burning bright light engulfed them' (2018, p 135). Stokes concludes that games are not simply governed by the logic of 'kill, or be killed'. Whether violent games increase the propensity to enact violence in the real world has simply not been determined. While it cannot be denied that some gamers have engaged in very hostile and threatening trolling, and some killers have enacted violent-game scenarios in real life, most do not; furthermore, games can promote pro-social empathy and cooperation. It is complicated.

5. Can the internet be regulated anyway?

Many believe digital networks make it impossible to prevent the circulation of pornographic content and that laws and moral norms have relaxed accordingly (Presdee, 2000). However, social, and legal systems can and do still uphold regulative forces (Hornle, 2011; see also Vera-Gray et al, 2021). Mike Presdee's (2000) *Cultural Criminology and the Carnival of Crime* identifies the emergence of a sadistic voyeurism within contemporary media and society. The commodification of this carnival of sadomasochism and hurt, including in the form of humiliation-based television formats and other kinds of quasi-pornographic representations, now pervades mainstream television, billboards and cinema, as these traditional media seek to compete with the ever increasingly explicit nature of sexual representations online. Presdee notes that demand for such content is fuelled by desire to escape the conformist and competitive pressures of an increasingly unfettered capitalism; but such material is not just a catharsis, he argues, as it also incites and normalizes a form of pleasure seeking based on the degradation of others. Transgression of rules becomes an industry of repressive desublimation (Marcuse, 1955), as carnivalesque reversal comes to reinforce hierarchies and profit.

The question of regulating such a culture is discussed by Vera-Gray et al (2021), in their examination of over 150,000 pornographic file names located on the three largest aggregator sites (Pornhub, XHampster and XVideos). Vera-Gray et al reject both 'effects' research and 'fantasy' (no-effects) models in favour of a 'scripts'-based approach. Scripts are said to socialize 'through stigmatizing and criminalizing some sexual behaviours while instructing and encouraging others' (2021, p 1244). Examining these titles, the researchers reject the 'myth of self-regulation' (p 1254), as, by their measure, around

12 per cent of 'titles described sexual activities that constituted sexual violence' (p 1249). The definition of 'sexual violence' used was broken down into four types: familial (incest); aggression and assault; image-based abuse (non-consensual or simulated non-consensual sexual images/revenge pornography); and cohesive/exploitative sexual acts. Enactment of familial sexual relations was the largest of the four types found, which raises the question of whether regulation of the taboo on incest, or its more likely simulation, incites the desire to watch such acts more than if the actors involved were not pretending to be related. That fabricating the transgression of a taboo appears to increase the popularity of viewing such acts challenges the notion that it is representation alone that stimulates desire. Attempts to regulate certain acts may also incite interest in precisely that which they seek to prohibit. The popularity of the term 'teen' (n=12,378) again raises the issue of transgressive ambiguity, and its use may indeed be designed to incite curiosity precisely because this age range crosses the boundary of legal regulation. Again, regulation may become part of how desires are scripted such that increased regulation stimulates rather than represses that which it seeks to prevent. Within the category of aggression and assault, once cleansed of erroneous cases, there were no cases of 'hit or hits', and only one case with 'rape' in the 150,000 titles analysed, while there were 703 cases of 'rough' and 830 cases of 'pound'. Of course, one case of 'rape' out of 150,000 is one too many. One can only hope it was removed when the researchers pointed it out. However, one case in 150,000 does not suggest pornographic aggregator sites are awash with simulated rape imagery. Similar differences were found within the categories of abuse imagery (voyeurism/-istic=902, hidden= 494, revenge=3), and for coercion/exploitation (with schoolgirl=756 and chloroform=4). The authors note that Pornhub has (since 2020) more actively sought to eliminate any such images of actual children when faced with a mass withdrawal of advertiser revenue, so it is possible to regulate, although where that leaves adults dressing up as schoolgirls raises again the question of transgression and the role of regulation in stimulating the desire to transgress. The four cases with chloroform in their titles, one hopes, have now been removed thanks to the research, and not gained increased interest due to publicity gained from media coverage of said research.

Attwood and Walters (2013) discuss E.L. James' *Fifty Shades of Grey*, a work of erotic fiction exploring, in graphic detail, consensual participation in sadomasochistic sex. The work became the UK's fastest selling book of all time, selling over 5 million copies in 2012 alone, and setting in train a trilogy of works given the appetite of readers to read more. Readers are almost exclusively female. The authors suggest:

> That a work such as *Fifty Shades*, emerging online as fan fiction,
> can become the fastest selling book of all time and yet individuals

are prosecuted for engaging with media texts that are already widely available, throws into question the way we understand media production, what we count as publicly available and accessible, and what counts as mainstream. (p 977)

That fictional representations of non-consensual sex may or may not be illegal in different jurisdictions, even as the conduct of consensual simulations of such acts may equally be legal or illegal in different countries, creates a regulatory space riddled with contradiction. Attwood and Walters (2013, p 977) conclude that 'the apparent drive to increasing regulation' within jurisdictions, 'alongside a striking failure to achieve successful prosecutions' because content flows across such boundaries, creates the impression of regulatory failure and subsequent demands for tougher actions. This tension has led to what Hornle (2011) calls a shift in emphasis towards 'inchoate crimes', where harm is assumed to be the future consequence of an act, even prior to any harm actually having occurred. This is manifested in the shift from criminalizing the production and distribution of illegal images towards the possession of images that depict or simulate criminal acts. Because it is possible to prosecute within a jurisdiction someone in possession of such content, possession is criminalized, whether or not it can be shown that possession in itself has led to the performance of any actual harm to others. Hornle suggests that existing legislation has actually been relatively successful in the prosecution of actual harm in the UK, and that actions by a range of state and non-state stakeholders (the Internet Watch Foundation, internet service providers [ISPs], and new-media platforms and users) have been relatively successful too. However, the desire to police public morals has tended to overshoot the need to prevent identifiable harm, and this, Hornle believes, has created a dangerous tendency to prohibit expression in the name of a generalized claim to be protecting society from immorality. Akdeniz (2010) argues similarly that the use of site-blocking tactics, whereby states instruct ISPs to prevent users in their jurisdictions accessing certain sites, can successfully regulate internet use (to a degree, given that VPNs can enable evasion for those actively seeking certain content). However, Akdeniz argues, this regulative capacity may, in fact, breach human rights and freedoms of expression. Those seeking tighter moral regulation tend to argue that regulation is failing, even as those who are more concerned to protect freedom of expression suggest regulation is already too successful.

In conclusion

The internet has made access to pornography – and increasingly graphic pornography – easier and more widespread than ever before. The capacity to access, to reach materials hosted in jurisdictions other than where a viewer might be located, and hence for evasion, becomes ever more feasible – that is, in relation to prosecution in one jurisdiction for content accessed there but which was not illegal in the jurisdiction where it was produced or distributed from. Nevertheless, scope for national jurisdictions to require ISPs to block content does limit such access/evasion within their legal space, even if VPNs allow evasions to occur by those already motivated to access such content. Most people do not use VPNs, but states with the highest use of website blocking also have the highest number of VPN users, suggesting where people want things, they can make the effort to get them. Also, a shift in legislation, from production and distribution towards the criminalization of possession, also reduces scope for evasion. Where the internet may give the impression of affording privacy in the consumption of explicit sexual content, such concealment is limited, as most users' privacy settings are not very high relative to what authorities might choose to deploy in relation to consumer possession. This limited concealment also applies to those who seek to share non-consensual ('revenge'/'abuse imagery'). However, the most significant issue in relation to pornography online is the question of incitement, which remains a source of heated dispute between those who believe explicit images can and do incite violent emulation and those who counter this claim. From those that advance media-effects accounts of causal links between violent sexual imagery and aggressive/abusing behaviour, to those advancing 'scripts' as a more subtle mode of corruption, and finally to those who wish to distinguish fantasy from any necessary relationship with real-world actions and beliefs, no consensus has emerged. Whether regulation itself actively incites desire for transgression adds another dimension to this unresolved domain of moral discourse.

5

Child Abuse Imagery, Abuse and Grooming

Key questions

1. Does increased circulation of abusive content increase risk to children?
2. Does grooming online extend real-world abuse or does distance reduce risk?
3. Do digital networks increase the scope for policing child sexual abuse online or reduce it?
4. What are the particular dangers offered by the internet to adolescents, those in transition between childhood and adulthood?
5. Do we know more or less than before about those who abuse and/or who consume abusive content?

Links to affordances

The issue of access in relation to child abuse hinges around the extent to which increased digital communication escalates risk when most child sexual abuse occurs within local networks, but where sexual images can be made and circulated remotely. The question of incitement is whether increased access to abusive content increases levels of physical harm or increases the extent of harm caused when an abusive act is circulated more widely. While increased circulation of content does not likely increase the overall number of abusive acts, the likely harm caused by such acts can be increased by wider circulation of those images. Regarding concealment, meanwhile, digital policing and multi-actor and international network policing can draw on the same resources as abusers, and so identification of online producers and consumers is perhaps easier than identifying actors engaged in real-world networks. However, revelation does not always make it possible to prosecute. As such, evasion may remain even when concealment does not, although legal harmonization reduces this problem.

Synopsis

While it is agreed that digital networks increase the scope to circulate abusive images online, even if levels of concealment and evasion are not as high as might commonly be assumed, dispute hinges again over the question of incitement. This is at a number of levels. Does the increased circulation of child abuse imagery increase the production of such content, which is intrinsically harmful to those children made to participate in it? Might increased circulation reduce overall production? Does the increased availability of such content increase the market for such material, and if so, does this increase the likelihood of consumers becoming direct abusers themselves? Even if watching child pornography does not 'directly' cause a person to become an abuser, does wider availability 'normalize' such action and so disinhibit action by persons already predisposed?

It is worth noting that online networks challenge national legal regulations, as laws regarding the age of consent differ between countries. However, it is also the case that the affordances of digital networks have led to international conventions on the age of consent regarding appearance in pornography that are more harmonized than laws regarding the sexual age of consent itself. As such, networks are not less ordered than the states they intersect. Legal harmonization around the criminalization of possession of abusive content also limits scope for jurisdictional evasion.

Similarly, while 'sexting' and 'livestreaming' of explicit content represents a new form of abusive communications from, to and between non-adults, where content is recorded this makes detection and prosecution easier than it is in relation to abusive verbal communication and physical behaviour. It is easier to expose and regulate such interaction because it is digital. However, not all digital content is recorded by service providers. WhatsApp may not record users' content if users delete it, but a recipient can do so. Most child sexual abuse occurs in localized and mainly familial contexts, not being established online. Recent exposés of child sexual exploitation have started with online disclosures, so digital networks may actually afford greater protection than real-world institutions did before young people had the means to speak out via digital networks.

Chapter sections

1. What is the relationship between abusive content and its consumption? This question returns us to the issue of media effects, and the wider issue of incitement. While researchers have offered typologies of those involved in child-sexual-abuse imagery, from consumption to production, the question of whether consumption incites or reduces further abuse is contested; and there is the argument that increased

circulation may reduce production (see Potter and Potter, 2001; Bentley et al, 2020).

2. Digital networks enable remote contact, and this may put children in touch with those who would seek to physically contact and abuse them. However, concern over remote communication should not distract attention from the primary risk that comes from those in immediate contact with children at home or in their immediate proximity (see Craven, Brown and Gilchrist, 2007; Lanning, 2018). Digital grooming is also easier to collect evidence on than real-world grooming communications.

3. Concern has been expressed that online communication and the circulation of abusive content online are harder to regulate in a global world, even if levels of concealment are not as high as many people (perpetrators and publics alike) believe. Nonetheless, significant challenges to traditional (patch) policing arise from network-enabled abuse (see Jewkes and Andrews, 2005). However, policing has not gone away in a global world, and new forms of international and multi-agency policing strategies and agreements enable its efficacy (Yar, 2013).

4. While the sexual abuse of young children is commonly agreed to be by far the greatest form of abuse possible, the incidence of harm to adolescents is higher. The question of how best to protect adolescents who are themselves beginning to explore their sexualities, and so sometimes unknowingly pursuing courses of actions that put themselves at risk, is of particular concern (see Marcum, Ricketts and Higgins, 2010; Halder and Jaishankar, 2013).

5. Media representation and criminal convictions create a picture of abusers that is unrepresentative and therefore a misleading guide to policy and to understanding individual actions (Reijnen et al, 2009).

1. What is the relationship between abusive content and its consumption?

David Wall (2007, p 109) suggests: 'The most contentious type of pornography is that which contains "sexualised or sexual pictures of children" (Taylor, 1999).' In one sense, this is incorrect, as child pornography is perhaps the domain around which there is greatest consensus. While disagreements exist, there is a greater consensus that child pornography is wrong than over any other area of law. Where the definition of a child is someone yet unable to make fully informed decisions about their own actions, children cannot give sexual consent and, as such, any sexual act involving a child is a crime. By extension, any recording and distribution of such acts cannot be consented to. However, if, by 'contentious', Wall means 'vilified', he is certainly correct. Nevertheless, digital networks have raised significant new affordances that impact upon the scale and potential harm done by child pornography, which in essence is child-sexual-abuse imagery. The relationship between such affordances and the harm done in the making of any such content is, however, debated.

Krone (2004) illustrates the complex nature of child abuse imagery in relation to law, suggesting a nine-point spectrum of offending: browsers; private fantasists; trawlers; non-secure collectors; secure collectors; groomers; physical abusers; producers of recorded acts of child abuse; and distributors of such content. These offences are not the same, but cannot be fully separated out within an overall ecosystem of online child-sexual-abuse imagery. It is one thing to say that browsing for such content is not the same as making and/ or distributing it. However, this point does not invalidate the counterpoint that browsing for such content may play a part in encouraging future production. To engage at the browsing level may or may not cause a person to be more likely to progress up the spectrum of abuse. Non-contact-based browsing may even reduce an individual's likelihood to progress to physical-contact-based abuse, but, in other cases, browsing may act as a gateway. Yet, despite this ambiguity, if browsing creates a perceived demand for content, such action may fuel future production and distribution.

Wall (2007, p 109) notes that, in the UK, the Protection of Children Act 1978 prohibited production, distribution and publishing of sexually explicit images of children, as well as possession of such content if it could be shown that such possession was for the purpose of distribution or display. It was only a decade later that possession alone was criminalized in the UK. Until very recently, Japan had not criminalized possession of child pornography, on the basis that production and distribution represented the source of harm while possession was seen as a passive consequence. However, in 2014, Japan did criminalize such possession, on the grounds that it incentivized abuse in encouraging further child-abuse-image production. For a long time, Japan was seen to be an outlier in relation to other highly developed economies in this regard. However, Japan's harmonization of its law here is interesting precisely because it illustrates that the global nature of the internet does not place digital networks beyond law: states have shown themselves increasingly willing and able to harmonize their laws so as to limit the scope for evasion.

Regarding harmonization, again, there is greater unity on the question of when it is legal to appear in a pornographic film than exists on the question of when it is legally possible to give consent to have sex per se. Where the age of consent to have heterosexual sex was 13 in England and Wales until 1885, after which time it was raised to 16, other countries in Europe have different laws, ranging from 13 to 18, and beyond Europe there is similar variety. Regarding the age at which a person can consent to appear in a pornographic film (or be photographed), some countries retain the principle that this should accord with the general age of consent, such that if a person is deemed able to decide to have sex, that person should be deemed able to decide whether that act can be recorded. Other countries, however, take the view that a higher level of consent should be required. Whether the harms that might come from sex itself are greater or lesser than those harms that might arise from its recording is therefore disputed. Yet, over time there has been a growing harmonization of laws in different countries around the age of 18 as the age of consent to appear in pornography, even if the age at which one can consent to having sex is lower. The law in Germany, for example, states that a 14-year-old can have sex legally, but not appear in pornography until 18. Estonia, meanwhile, has the same age of consent (14) for both acts. By German standards, an Estonian 16-year-old appearing in pornography is being abused. Should German law extend to Estonia? In 2021, Estonia's parliament debated raising the age of consent for appearance in pornography to 16. Should German laws apply to Estonian teenagers? If Estonian lawmakers do not think so, is that something that Germans should be worried about? If a UK citizen were found in possession of such images of a 15-year-old Estonian, they would be subject to UK law, as would an Estonian in the UK. A UK citizen would be subject to UK law even if they committed an act in Estonia that was lawful in Estonia, if that act could be proven after their return to the UK. Whether UK law should apply to Estonians in Estonia is another matter. Those who support the view that 18 should be a universal standard point to countries that do not apply that standard, and this has been successful in changing laws in many countries. Australia, for example, has national law making the age of consent 16 for both having sex and appearing in pornography. This can be shown as evidence that a lack of harmony means sexually explicit images of 16-year-olds (a criminal offence to produce, distribute and possess in the UK) are accessible to UK internet users via Australian sites. However, many Australian states have recently passed state-level regulations that raise the age of consent for appearance in pornography to either 17 or 18, in line with international campaigner standards. As such, once again, the argument that the internet undoes regulation is untrue. International harmonization in relation to child pornography is, in fact, relatively strong.

However, this is not to suggest that there are not significant dangers and new affordances for harm arising from digital networks. The UK's NSPCC (National Society for the Prevention of Cruelty to Children) highlights new and emerging trends in online abuse and child abuse images online (Bentley et al, 2020). Where two generations ago, children's access to online networks was largely mediated by schools and libraries (Potter and Potter, 2001), and then, a generation later, by computers at home (which offered at least some level of parental oversight and control), today, in the UK, 70 per cent of those aged between 12 and 15 have their own mobile phone with access to the internet (Bentley et al, 2020). This creates a space where adolescents are required to self-regulate, and often this fails. Internet Watch

Foundation research cited by Bentley et al suggests that, between 2017 and 2018, the number of URLs containing sexual images likely to be of adolescents almost doubled (although the scale of increase in the following 12 months was less sharp, it was still increasing). In part, this increase is due to greater awareness and reporting, but another significant factor was the rapid increase in self-posted content ('sexting') by under-18s. Most sexually explicit images of under-18s online are of teenagers, not pre-teenagers, and around half of such sexually explicit images of teenagers are self-posted. Here, the traditional distinction between the physical abuse involved in producing child pornography and the harm that can arise from its circulation is reversed, as the greater harm in relation to self-posted content is in its circulation. However, sentencing guidelines that distinguish between images that contain explicit posing and those that contain sexual contact make it possible for laws that already exist to distinguish different harms and to prosecute accordingly. While self-posting or 'sexting' is a new and harmful affordance of mobile-digital-network technologies, it is not beyond the law to address it in a proportionate fashion.

Controversy over fabricated images of child abuse (such as computer-generated animations) highlights the problem of identifying harm beyond the act of physical abuse that occurs in the making of child pornography. Ost (2010) argues that an animated/computer-simulated representation of a child being shown engaging in sexual acts cannot be said to involve harm to any actual child; as such, the legitimacy of the UK Coroners and Justice Act 2009, which criminalized the production, distribution and possession of such images, is, she claims, disputable. Ost argues (2010, p 232) that a liberal conception of identifiable harm as the basis for prohibiting any action does not warrant criminalization in such cases. However, this liberal conception of harm is not one that everyone agrees with, and even liberals cannot agree when a generalized insult to a group becomes a threat. If a specific child may be said to be degraded by the circulation of an abusive image of that child, why should it not be the case that children in general might be degraded by abusive images of figures that are clearly meant to be children? Ost is concerned to limit the idea of a general morality in need of protection, relative to a harms-based principle that individuals should be protected from specific threats. She may or may not be correct in this belief. That any particular society might seek to uphold its own 'general' morals over and above the protection of specific individuals from specific harms (such as, for example, banning offensive cartoons that caricature particular ethnic, racial or religious groups even while not explicitly calling on people to attack members of such groups) may or may not be correct. What is true, however, is that despite the global nature of the internet, the ability of states to legislate remains possible. Access to children has increased, even if overall or specific additional harm is harder to quantify as what might incite harmful behaviour in certain viewers, might have a cathartic (or no) impact on others.

2. Digital networks, grooming and child sexual abuse

Whether the incidence and risk of child sexual abuse has increased as a result of the internet is disputed. What is not in dispute is that concern over child sexual abuse and grooming has increased alongside the rise of digital network technologies. Concern over online access to children has certainly increased attention on the issue of grooming, that is, communication with children with the intent to commit acts of sexual abuse.

Kenneth Lanning (2018) sets out to document the history of the term 'grooming' in its relationship to the policing of child sexual abuse in the United States. In the early 1970s, the primary focus of legal and policing concern regarding the sexual abuse of children was in relation to 'stranger danger', and the focus was, therefore, on force, threat and abduction prevention, awareness and investigation. What emerged in the course of that decade was a dual refocusing: first, upon domestic child sexual abuse by family members; and second, upon 'acquaintance' abuse, by which child molesters gained access to victims through means other than coercion (befriending, attention, affection, gifts, money and privileges). Such methods of gaining access to and control over children mimic forms of befriending and relationship formation that may be considered normal between peers (persons of the same age). The term seduction, Lemming notes, came to be used to refer to such means of access and control between an adult and a child for the purpose of sexual abuse. In part, the term seduction captured both the sexual dynamic of such relations (which, if occurring in relations between peers, might be accepted as unproblematic) and, importantly, the sense of the inappropriateness of this dynamic given that the relations are, in fact, between non-peers (between an adult and a child). However, in the 1980s, the term grooming came to replace seduction in reference to such sexually oriented communications (see Groth and Birnbaum, 1979; Conte, 1984, for examples of this migration in language).

Lanning (2018, p 11) writes: 'I define grooming/seduction as the use of non-violent techniques by one person to gain sexual access to and control over potential and actual child victims.' Focus upon the non-violent mode of control is significant, as Lamming notes that this means of gaining 'compliance' from victims is what creates the false impression of 'consent', when a child, by definition, is not capable of giving such consent. Grooming creates conditions where a child complies with their abuser. It is essential that such abuse is understood as being what it is, and grooming the means by which such control is achieved. Where violence is not a factor, documenting the abuser's mode of communication is essential in understanding how the abuse was achieved and in evidencing it in court. In this respect, the rise of online grooming has increased the capacity to identify abuse, and may, therefore, have made it easier to successfully prevent it. Lanning notes (2018, p 13) an issue that was and remains pressing: 'For youth-serving organisations, most problematic is the difficulty in distinguishing grooming from mentoring.' If those seeking to gain access and control over children in order to commit sexual offences actively seek to involve themselves in organizations that provide child-oriented services and activities, participating in such organizations creates opportunities for interaction with children that are considered to constitute legitimate mentoring (such as in sports coaching) and these interactions are often unsupervised. In the 21st century, online forums and spaces have replaced or supplemented some of these organizations. Against this backdrop, it may not be that access to children by non-familial members has increased per se but, rather, that the means of such access have grown. Ironically, online interaction is perhaps more readily identifiable and may, therefore, make detection and intervention easier than it was in the past. The digital does afford new modes of access, but levels of concealment and evasion by such means may be reduced relative to older forms of acquaintance-based access.

In addition to the risk from acquaintance-based, face-to-face abusers, compared to that posed by online strangers, there is also the relative risk of abuse from family members and strangers. Craven, Brown and Gilchrist (2007) cite the work of Carr (2004), who identified 27 cases of child sexual abuse over a two-year period in Britain having been facilitated by online communication. This is, of course, likely to be an underestimate of the true extent of such online grooming. However, Craven et al note that, even if it is an underestimate, such abuse by strangers facilitated by the internet is very small when compared to estimates from the NSPCC that suggest as many as one in ten children experience some form of sexual abuse from a member of their own family. While online access is a real and significant threat, it should be noted that face-to-face grooming and abuse are by far the greater threat, and that access to online communication may be a way of highlighting familial abuse. While concern about online predators is often linked to calls for parents and guardians to be better informed about how to control and limit their children's access to and communication on digital platforms, such control may itself become a form of grooming if children that are being abused in a domestic context are manipulated into compliance and concealment of their own abuse by means of familial 'protection'. Online communication beyond familial guardianship may be the best way of raising concerns.

Craven, Brown and Gilchrist (2007) outline the emergence of the Sexual Offences Act 2003 in England and Wales, and, in particular, its

Section 15, which deals with online grooming. In 2001, Patrick Green was arrested and charged after arranging online to meet a 13-year-old girl. Green had claimed to be a 15-year-old boy. The girl's mother, being concerned with these online communications, had followed her daughter and intervened when she found Green to be a middle-aged man. Green was arrested and charged, but he was not convicted of any crime as the law at that time did not specifically outlaw communication with a child if no physical sexual offence had yet been carried out. At that point in time, the Criminal Attempts Act 1981 did make preparation to commit a crime an additional crime if it were undertaken alongside other criminal acts – such as the unlawful possession of a weapon. In the Green case, preparation to meet a child for a sexual purpose was not deemed technically an offence if no abuse had actually yet occurred. Likewise, the Child Abduction Act 1984 could have been used, but as the mother had intervened before any physical act had taken place, abduction had not 'happened'. As such, the Sexual Offences Act 2003 criminalized grooming, as communication with the intent of meeting a child for the purpose of committing sexual abuse. Craven, Brown and Gilchrist are highly critical of the limits they see as existing within the 2003 Act, but what is perhaps most important here is that they identify that such problems with the Act relate mainly to the fact that online grooming is so much easier to identify than face-to-face grooming and abuse. These authors are rightly concerned that focusing attention on online grooming distracts attention from far greater and far less visible forms of grooming in real-world situations. That online communications are more readily retrievable than face-to-face communications is one thing. That online grooming requires travel to actually meet a prospective victim also increases the scope to prove 'intent' relative to meetings that occur in either the victim's or the perpetrator's routine locations.

However, while online communications are more likely recorded and retrievable than face-to-face communications, this is not simple and straightforward. While platforms like WhatsApp have age restrictions, such that under-16s are not allowed to use the platform, the service does not use age verification to confirm user age, so a child can falsify their age – although, on the flip side, a person seeking to groom children on such a service would not be able to falsify their age if that meant claiming they were themselves a child. Police forces and informal vigilante actors can similarly pose as children on forums that do allow children access (such as certain games platforms) in order to entrap sexual predators, such that the ability to conceal real age online cuts both ways. Digital networks do afford access to children by remote groomers, but it should be recalled that most such predators are proximal familiars, such that giving children means of communicating beyond their familiar domains may increase safety in some cases.

3. Is abusive content online harder to regulate in a global world?

'These abusive images of children were like sunken wrecks filled with dirty oil lying at the bottom of the sea – they were out there but nobody knew they existed. Then the Net came along and it allowed all this filthy stuff to bubble to the surface' (Detective, cited in Jewkes and Andrews, 2005, p 42). The internet has certainly made child-sexual-abuse imagery more readily accessible, even while the increased accessibility of such content has shifted the ground in terms of concealment and evasion. Jewkes and Andrews (pp 44–45) go on to quote another police officer on the question of online evidence in relation to child sexual abuse:

> 'One of the best things about our work is that this is one situation where the defendants cannot accuse the police of putting words into their mouths ... It's all right there on their hard disk, every word of it. It is always clear that the issue of sex is introduced by them, not us. It's irrefutable.'

In the early 2000s, it was necessary to upgrade legislation in many countries to make digital evidence admissible in line with 'physical' evidence; once done, however, this has meant 'the problem' is not a lack of evidence, but, on the contrary, the sheer scale of the evidence that has now become available. Jewkes and Andrews note that there are now (and what they argued here in 2005 remains the case today) problems with connecting between forces across jurisdictions, as well as in terms of gaining sufficient resources to process the scale of evidence being made available by online means, and in terms of updating the knowledge, training, skills and orientation of police forces in relation to digital content. Where traditional policing tended to be localized and driven by victim reporting, digital evidence now starts at the non-local (network) level, and also now often starts with perpetrators being identified rather than victims coming forward. Where, up until the early 2000s, the primary source of incriminating evidence came from networks sharing child-sex-abuse imagery based on central servers distributing content to subscribers, this has shifted now towards peer-to-peer distributed networks. Where these earlier models allowed the police to capture large numbers of individual users' credit card details, prosecutions for possession alone were often hard to achieve. In addition, it tended to be hard to prosecute the suppliers if they operated servers in less well-regulated jurisdictions (thus a means of evasion). This explains the shift in legislation towards making possession alone into a crime.

Krone (2005) examined 31 police-led, anti-child-sexual-abuse-imagery operations. He notes the radical rise in concern over child sexual abuse in the decades leading up to the early 21st century. This trend in media concern has continued apace in the decades since 2005. Krone suggests there to be four types of police operation, which are as follows: 1. those targeting individual uploaders; 2. those targeting covert groups engaged in creation and circulation of abusive content; 3. those targeting website-subscriber service providers; and 4. police sting operations targeting individual content browsers. Individual uploaders who make what they are posting available outside of a closed circle are creating content and making it accessible in a way that it might readily be seen by those not already actively seeking it. This may be seen as doubly dangerous, in relation to those abused in the production and wide circulation of such content and in terms of corrupting those who might not already be seeking it but who might come across it. However, such distributors, by the very extent to which they make content accessible, also make themselves more vulnerable to being identified. Covert groups maintain different levels of closure. The more closed they are, the greater the levels of concealment and evasion such groups can maintain, and the narrower is the scale of access to outsiders. However, the wider such groups make access to the content they produce/distribute, the greater the risk they face in being identified. Website subscription services, in taking the credit card details of users, make such users vulnerable if the service is itself accessed by authorities; and, as such, these services have declined in relation to the rise of online file-sharing services, which themselves then succumb to the risks as identified in relation to relatively open user groups. Police use of sting operations, where content is advertised in order to lure those wishing to view it to identify themselves, again highlights the cat-and-mouse nature of access and identification (the breaching of concealment). A criticism of policing in the early years of the 21st century was that an increasing number of identifiable cases of individuals accessing indecent images of children online was not resulting in a comparable increase in the number of convictions. This issue hinged on generating large numbers of identifiable cases of access within particular jurisdictions, but far fewer identifiable cases linked to the production and supply of such content. As such, the issue of evasion (and, to a degree, concealment also) was significant. This has, however, been addressed in the years since Jewkes and Andrews, as well as Krone, identified the problem, through making possession of child abuse images alone a sufficient condition for criminal conviction, as mentioned earlier. As such, it is now the case that local jurisdictions can successfully target individuals in their 'patch' even if content has come from somewhere else (or where its original source cannot be easily identified). To an extent, access can be blocked at a local level, and where access may be sufficient to secure a conviction (even when the content was covertly supplied as part of a police sting operation), it is mistaken to argue that global networks render local policing powerless. They do not.

Majid Yar (2013) highlights today's increased dispersal and pluralization of policing. The term 'policing the internet' means many things in relation to many actors, from ISPs and platform hosts to user groups and non-governmental monitoring services, as well as state-funded, local, national and international police forces. However, Yar argues, it is also true that state-funded policing is capable of focusing resources on child-sexual-abuse-related content as a matter of priority. Police forces in the UK (and, in parallel, in other jurisdictions) operate 'hierarchies of standing', by which crimes are ranked in terms of the seriousness of the threat posed and the relative vulnerability of the potential and actual victim. Online sexual abuse has become a priority within such hierarchies of standing in recent years, as have sexual offences more widely. Within the general field of online sexual abuse crimes, self-victimization content crimes (such as sexting by children; see Bryce, 2009) have seen attention from service providers, non-governmental monitoring groups and formal state police forces alike. However, adult-to-child-related abuse (in terms of both grooming and explicit content) has been seen as more serious and, therefore, as a priority for state-funded police action. However, where resources are limited, to prioritize adult abuse of children over self-victimization – while rational within a 'hierarchy of standing' – does not cancel out the question of whether more resources should be deployed. However, while questions of resourcing are significant, this is only true because such resources are able to make a difference. Even while policing tends to remain geographically limited, it is not ineffectual in a world of global networks. Concealment can be amplified online, but the digital, by its nature, always leaves a record in a way that 'real life' may not; and while evasion by distributed networks and remote access has proven hard to regulate within any particular jurisdiction, it is precisely in confronting this difficulty that harmonization of legal frameworks and police cooperation has moved forward apace in recent years.

4. Self-generated content and the protection of young people from themselves

Sexual abuse of young children is commonly agreed to be the greater abuse relative to older victims. However, the number of adolescents harmed is higher. How is it best achieved to protect adolescents as they begin to explore sex and sexualities, and so sometimes unknowingly undertake actions that increase their risk (see Marcum, Ricketts and Higgins, 2010; Halder and Jaishankar, 2013)? As was noted earlier in this chapter, the work of Bentley et al (2020) suggests that around half of the sexually explicit content online of those over 13 but under the age of 18 is self-generated and self-uploaded. Halder and Jaishankar (2013) compare the legal responses of the American and Indian legal systems in attempting to deal with such self-generated, but illegal, content. Any content containing explicit images of under-18s created or uploaded by an adult would constitute child pornography in both countries and subject to a long prison sentence. When the parties involved in making and uploading such content are themselves children, the law must take a different approach, but what approach exactly this is, is complex and has to balance competing harms. Where content has been posted by a child who is subsequently victim to that material being more widely circulated by another individual under the age of 18, the motive of the person circulating the content is not usually profit, but rather the intent to take revenge on the victim for a perceived slight (such as in ending a relationship). An adult intending to profit from such content would be classed as a child pornographer, but an under-18-year-old, acting out of revenge, would not. However, intent to discredit and humiliate someone by breaching their privacy is also a serious crime. Can a child be said to have the capacity to hold such criminal intent? Successful prosecutions of under-18s have been undertaken in the United States, even while emphasis has been more towards service providers and removing service providers' protection that might arise through 'camouflage pornography' (whereby material is presented as the victim's own free expression). Platform hosts have an obligation to remove or block content that is not authentically consensually produced, but this still leaves those uploading material subject to prosecution; and if half of these unloaders are themselves minors, the question of whether this is the right approach remains. In India, attempts to deal with revenge pornography are more limited, and Halder and Jaishankar note that the most widespread response has been the wholesale prohibition of the use of social network sites and mobile phones by children within school contexts. This effectively removes any duty of care from state institutions concerning the supervision of children's own online actions, protecting state institutions. Whether such prohibition on school premises actually protects children is another matter. The alternative, which would be to enact protections for children on supervised machines (school and library computers) and in the use of children's own machines (mobile devices) while at school, creates a whole new level of difficulty and liability, which such agencies (as legal guardians) may be insufficiently prepared to successfully undertake. Halder and Jaishankar suggest what they call a 'therapeutic jurisprudence' approach, which seeks to care for – rather than punish – children for engagement with digital technologies in the conduct of emergent and natural (if often problematic and distressing) sexual exploration and development. Where legal prosecution may label the perpetrator a sex offender while still only a child, prosecution may also raise the profile of the very harmful content that victims suffer from others knowing about.

Marcum, Ricketts and Higgins (2010) address the question of how best to protect under-18s from themselves, adopting a routine activity theory approach to risk and harm. In a study of teenage digital technology use, Marcum, Ricketts and Higgins found that increased use of digital technologies increased the risk of accessing harmful content and communications; and, as children age into their teenage years, exploratory engagement with sexually explicit materials increases, along with the risk of exposure to harmful content and communications. This may be rather predictable, and it is the next part of Marcum, Ricketts and Higgin' research that is perhaps more important. What the authors found was that levels of protection software on the machines used by teenagers had no impact on the level of exposure to harmful content and communications. In essence, a belief in the value of such technical guardianship is misplaced. What the research found to be effective was the presence of real-world 'guardians' in the spaces where young people are accessing and using digital services. If a young person feels/knows/sees that they are under supervision, they are much less likely to access and engage with sites and content that are risky. This suggests that such teenagers are in fact capable of knowing, to some extent at least, what it is that guardians would not want them to access, and, when in the presence of such actors, tend to avoid it. When only protected by technical guardian software, teenagers are less likely to enact the levels of restraint they do undertake when they know they are being watched. On the one hand, this research suggests that teenagers have a significant capacity to know what guardians would not approve of, and, as such, they would be better able to protect themselves than guardian software; but the problem is that these same teenagers are often unwilling to abide by such restrictions, and so still require supervision. Whether schools, parents and other responsible adults are willing to undertake this role, and the extent to which teenagers will actively abide by this need to be supervised, is another matter. Still, it is the case that protection can be offered. Technology alone is neither capable of supplying protection nor of preventing it. Real-world surveillance reduces self-concealing, evasive access on digital actions by teenagers better than any affordances of the digital as such.

5. Producers and consumers

News bias and conviction rates are unrepresentative. These then become a misleading guide to policy and for understanding individual actions (Reijnen et al, 2009). Yar and Steinmetz (2019, p 175) note that high-profile media personalities, when caught in possession of child abuse images, gain a disproportionate amount of attention in a mass media already fixated on such celebrities. They cite the examples of the UK musicians Gary Glitter (Paul Gadd) and Pete Townsend, as well as the US sports coach Larry Nasser. More recent cases of high-profile media, business and political figures convicted of sex offences involving children has added to the impression that abusers are more likely to be a certain kind of socially successful person, when the opposite may be closer to the truth. Adams and Flynn (2017) carried out an analysis of internet-related child sex offences committed in the United States between 2004 and 2013. Those convicted were disproportionately male, and predominantly white men. While the population of the United States is itself disproportionately white, it is the gender composition of these results that is most starkly clear. Reijnen et al (2009) identified significant differences within a sample of offenders, some of whom had been convicted for offences related to the possession of child abuse images gained from the internet, and others who had been convicted of sexual offences with children (a control group of non-sexual offenders was also studied). What Reijnen et al found was that those convicted of content possession were disproportionately likely to live alone, have no partner and also have no children, relative to those convicted of contact-based sex offences against children, who were more likely to be cohabiting and have children. This research suggests that consumption offenders were predominantly social isolates, although whether criminally oriented sexual desires caused a tendency towards social isolation, or whether isolation caused a tendency towards deviant sexual desires, could not be established. In contrast, contact-level child sex offenders were much more socially integrated. While cause-and-effect relationships are hard to determine, it would appear that the kind of person that consumes child pornography is less likely to have the social relations or social skills required to access and abuse children. Less likely, however, is not the same as saying this connection cannot, in some cases, exist, or that, in some cases, the former offence may not incite the latter offence. What this research does show, however, is that conviction for child pornography possession is not a predictor of actual abuse, and it would be dangerous to assume that the biggest threat to children comes from isolates on the internet relative to more socially integrated individuals.

In conclusion

While access to sexually explicit material involving children has increased as a result of the spread of digital networks, concealment and evasion are not at all as clear-cut. The very nature of digital networks means that conduct that might once have remained private can circulate, but, at the same time, criminal communications and conduct/content that might once have remained hidden are now recorded and available to prosecutors. Distributed networks that traverse legal jurisdictions create problems for patch-based 'local' lawmakers and police, but, at the same time, recent decades have seen increasingly harmonized laws on such things as the age of consent to appear in pornography, so reducing scope for evasion. The claim that the digital increases levels of incitement to commit further offences against children, whether that means producing further abuse imagery or acting out such conduct after watching such content, remains disputed. Symmetry or asymmetry arise from contingent resourcing and hierarchies of standing, conflicts over digital evidencing of preparatory crimes versus pre-contact denial; as well as localized attributions of responsibility versus distributed liability (attributed to evasive and remote actors).

PART III

Corruptions of Citizenship

6

Privacy, Surveillance, Whistleblowers and Hacktivism

Key questions

1. Do digital networks afford more privacy or less, relative to pre-digital times, and for whom?
2. Can citizens trust their state's laws to protect data from global surveillance capitalist firms, or do these firms give citizens the means to learn of and resist state intrusion?
3. In what ways are hackers similar to and/or different from whistleblowers?
4. Does the term hacker conceal more than it reveals about the diversity of activities that are categorized under this label?
5. Is hacking now more an arm of inter-state power struggles (cyber warfare) than a form of resistance to the powerful (hacktivism)?

Links to affordances

Invasion of privacy requires the ability to gain access to other people's computers. The cat-and-mouse game between surveillance and encryption highlights that such access, and the ability to evade surveillance and retain the capacity to conceal content from such attempts at access, is never fully resolved. Successful intrusions by hacktivists have incited a backlash against state invasions of privacy, but, at the same time, self-exposure on social media continues, despite revelations about big-data abuses by surveillance capitalist platforms. It is possible to conceal (to a relatively secure level) a great deal more than most people care to do; whether those that care do so because they have something to hide should not be assumed. Ultimately, data abuses, when linked to micro-targeted marketing, allow some degree of behavioural management in the sale of products, while the ability to manipulate elections seems less powerful than often may be assumed.

Synopsis

Because the scope for privacy has expanded exponentially in recent years, and in particular in online spaces, what privacy there is to be invaded has also grown. The scope for privacy and its invasion is, then, a cat-and-mouse game between individuals and institutions, between institutions and between individuals. The scope for states to monitor individual communications is greater than ever before, as is the scale of big data collected by companies and now subject to ever more sophisticated forms of analysis for the purposes of profiling, micro marketing, behavioural prediction and choice modification. Yet, awareness of attempts to monitor citizens and manipulate consumers is also increased by means of digital networks. Such networks inform as well as inform upon us.

In what ways are WikiLeaks and Anonymous similar? In what ways are the two different? Is it possible for social scientists to have an objective view on whether leaks and hacktivism are 'justified', or is such judgement only ever a by-product of whose side you are on? Can social scientists be neutral or is it in the nature of social reality that facts are morally loaded? Do the cases of Chelsea Manning and Edward Snowden show the scope or the fragility of power in today's digital surveillance state?

Chapter sections

1. Today's states and corporations gather far more information about us than ever before. Concern about privacy has grown as a result, but such threats to privacy have also heightened demand for privacy. Desire for privacy has grown even as the scope to invade it has also grown (Nissenbaum, 2011).
2. States, corporations, citizens and social movement actors engage in competing invasions and assertions of privacy. such actors all claim to promote the public interest, but none truly defend it; even as each curb the power of the others to undermine it (see Deleuze, 1992; Zuboff, 1988, 2015 and 2019).
3. New digital whistleblowers reveal how networks afford breaches of privacy, often of powerful actors themselves engaged in infringement practices (Beckett and Ball, 2012; Bauman et al, 2014; Coleman, 2014).
4. Hacking has changed – from technical play to malicious harm and, finally, to politically motivated hacktivism and cyber warfare – not by a logical progression of any 'spirit of informationalism'. Information does not 'long' to be free. Play, vandalism and revelation are matters of interpretation; as is the question of the black, white or grey colour of a hacker's hat (see Steinmetz, 2016; Jordan, 2017).
5. Wall (2012) claimed the Russian hacker as folk-devil was a largely mythical (or highly exaggerated) threat. It is now suggested that this moral panic has

become a reality through Russian state backed cyber warfare (Determann and Guttenberg, 2014).

1. The parallel loss and gain of privacy in networks

Concern today about the loss of privacy to corporate and state actors, as well as to other individuals, is in part a reflection of greater risk; but it is also that we have come to demand and expect levels of privacy that were unimaginable in past times. Helen Nissenbaum has been most prominent in critiquing the standard assumptions about privacy governing the internet, principles that have extended from assumptions largely set in the United States. However, it should be noted that even in parts of the world that appear similar to the United States, in respect of attitudes to privacy and free expression, significant differences remain, and large parts of the world have very different legal assumptions. As such, both Nissenbaum's critique and the alternative model she offers may be themselves limited to a particular national context (the United States). However, the extent to which internet cultures reflect US patterns of action and interpretation may mean her arguments can be extended. In her book *Privacy in Context* (2010), and its later summary (2011), Nissenbaum sets out her critique of standard 'notice and consent' privacy policies. The core problem she identifies with this model is the 'transparency paradox'. She concludes by setting out her alternative, 'context-specific substantive norms', or 'contextual integrity', model. The standard 'notice and consent' model of privacy gives the user of a digital service a choice to sign up to the providers' 'terms and conditions', as drawn up by the service provider. This privacy policy gives notice to the prospective user of how their data will be used, and the user is then given the 'choice' of whether or not to accept these terms and conditions. The problem with this model, Nissenbaum argues, is that most users cannot comprehend the language used in such terms and conditions. Often, users lack alternative services, so 'choice', even if it were 'informed' (which the first problem as noted means it most likely is not), is also missing. The notion that the user has given informed consent to what they then 'click' acceptance of is therefore highly questionable. The model assumes that the service provider has made clear all the potential uses to which the service user's data will be put, as well as all the potential users of this data, but this raises the 'transparency paradox'. For all such uses and users to be set out and made clear, the terms and conditions being offered to the user would have to be very long and complex, meaning it is very unlikely that a user would have the capacity or the time required to engage with all such content fully. If, conversely, the uses and users were explained in a simplified fashion, much of the significance of such uses and users would be lost. That is, then, the transparency paradox. Acquisti et al (2015) found that while social media users have become more aware of the need to manage what is made public on social media platforms, it remains the case that most do not fully understand what is being made available about them when using such services, even having clicked on a button to 'agree' with a service's terms and conditions. Barth and de Jong (2017) identify what they call a 'privacy paradox' (akin to Nissenbaum's 'transparency paradox'), by which people express high concern about their privacy but do very little to protect it, often assuming contextual protections (safeguards) on the services they use that simply do not exist. People often want to believe they are being protected, even when there is little evidence to support such faith.

What, then, is the alternative to the 'take it or leave it'/'opt out' model, which gives control primarily to service providers relative to users? Nissenbaum suggests that context-specific privacy rules exist in many social contexts, such as between patients and doctors, and between a bank and its clients, a school and pupils/pupils' parents, and so on. Such principles have arisen over time, and are not simply devised by service providers such that they can then maximize additional benefits from selling on client data to alternative end-users. Banks, schools and hospitals, for instance, are governed by formal regulations and by informal expectations about what is appropriate. These rules could equally be applied by appropriate bodies seeking to regulate ISPs and social media platforms, by setting out privacy policies for such providers rather than allowing providers to simply offer prospective clients 'take it or leave it' options to sign up to conditions they (the provider) find most conducive to their business interests. That the European Union has sought to develop privacy rules that would apply to service providers illustrates this possible alternative approach, but also its limits (McDermott, 2017). Nissenbaum can note, however, some success in the take up of 'contextual integrity' as a model for privacy in some sections of the computing industry and network service provision (Benthall, Seda and Nissenbaum, 2017). Whitehouse (2010) has also sought to extend Nissenbaum's approach to the development of social media-platform ethics policies. Multiple privacy breaching scandals in recent years suggest this has been of only limited success, both in terms of the conduct of digital services and in whether (in principle) contextual norms can be formulated and agreed between competing interests and groups (Rule, 2019; see also Igo, 2018). States, whether they be the United States, India or China, assert the right to regulate online platforms in line with what they claim are their national societies' cultural contexts. Critics may then argue the relative merits of service providers or states when it comes to preserving the privacy of citizens. That the United States Government also intervenes to require its own service providers to store and make available data (to national security and police agencies) about its own citizens also raises the question of what 'context-specific substantive norms' apply when: those of citizens' rights to privacy or society's right to national security? Amoore (2014) questions whether the law surrounding the privacy of individual citizens can really protect citizens from the very states that are tasked with enforcing such laws. When such intrusions are revealed, citizen behaviour (in adopting stronger privacy software) may be more important than new laws. The public reaction to WikiLeaks' revelations is one case in point (Moore, 2011). Blanke (2018) likewise argues that, as privacy is a flexible term, it is only in the scale of outrage over intrusion that boundaries are established, entrenched or altered. As public concern over privacy fluctuates and is riddled with contradiction ('the privacy paradox' again), protest can even cause intrusion to expand (as when legal cases draw attention to that which they seek to conceal; see Jansen and Martin, 2015). Park (2013, 2015) found additional complexities and perhaps paradoxes in the differences that internet users displayed in relation to privacy, depending on age, marriage, gender and level of education. There is no simple gender pattern in terms of confidence, protection and personal data release, once age, marital status and education are taken into account. Users appear to be balancing multiple competing constraints, interests and levels of awareness, when seeking (successfully or otherwise) to protect their privacy while communicating online. Davis and James (2013) found that 'tweens' (their sample being aged between 10 and 14) display a complex balance of awareness and concern for privacy, even as they are not always fully aware of risks and their privacy protection strategies remain 'haphazard' (which makes them rather more similar to, than different from, adults).

The idea of privacy itself emerged alongside the very development of new means of invading such private space (Baghai, 2012). Emile Durkheim suggests that 'moral individualism' emerged as a particular 'collective representation' during urbanization and industrialization (Durkheim, 1975). The space for private action and choice grew precisely in line with the intensification of social integration, greater interdependence and proximity to strangers. Similarly, Georg Simmel (1906) suggests that the transformation from secrecy to privacy occurs with the creation of increasingly urban living in which larger crowds exist alongside increasingly private living arrangements. Baghai notes also Norbert Elias' (1994) 'civilizing process', by which the realm of private bodily functions becomes increasingly separated from public view, and that ideals of decency and discretion, once reserved for elites, come to be seen as common courtesies. Erving Goffman's (1952) front and back stages become the management devices for all – so much more traumatic, then, when they are breached, whether in an institution, on the street or along the boundary of private and public life. In this vein, Gangadharam (2015) found that in the rush to connect socially excluded groups in the United States, outreach and training programmes designed to get the poor connected tended to be 'privacy-poor, surveillance-rich' systems, such that 'inclusion' was occurring in very many negative (as well as positive) ways. Yet, as newly included groups are incorporated into older forms of digital surveillance, so it is that the Internet of Things offers to generate surveillance in new ways for those able to afford newly integrated, network-connected objects: from cars, fridges and homes, to fitness devices and personal organizers (Maple, 2017).

The essence of Baghai's argument flows from the work of Niklas Luhmann's (1996) theory of increasingly differentiated subsystem specialization. As society becomes increasingly organized into relatively self-regulating, functionally differentiated systems and subsystems (governed by their own rules, goals and role allocations), so privacy becomes one key dimension of limiting the intrusion of one subsystem into the functioning of another, especially when human agents operate within multiple subsystems on a daily basis. Larger and larger parts of a person's life become of no concern to the subsystems outside of which such actions take place. The rise of digital networks create communication between systems and subsystems that breach both the norms that Nissenbaum discusses and systems' self-regulating exclusions that Luhmann claims are key to enabling complex societies not to collapse into noise (noise being non-relevant communication between subsystems). Failure to control such non-relevant communication, for Luhmann, creates noise that itself fuels risk anxiety (which Luhmann's pupil Ulrich Beck developed as an idea). Such noise can also be understood as breaching privacy if information from one sphere passes to another sphere and, in so doing, disrupts the functional integrity of the former. Vitak (2012) refers to this as 'context collapse'. Digital networks, by the nature of their distributed structure, afford significant levels of such breaching/context collapse. Still, it should be remembered that demands for privacy emerge from the same processes that challenge it, and digital networks afford certain kinds of private communication, the breaching of which is taken to be outrageous precisely because many users have come to expect a level of privacy that might have been hard to maintain in the past. Agger's (2015) concept of 'oversharing' reflects the tension between what people are willing to share online with more limited revelations in other contexts. Agger's work suggests that Nissenbaum's contextually established limits are readily breached in emerging digital spaces because users allow them to be; but Agger also suggests that recreating boundaries is a feasible project of individuals and societies. Both agree that technology does not compel the collapse of privacy regulating social frameworks, even if each focuses on a different dimension of what does (surveillance based corporate profit motives for Nissenbaum, and individual willingness to overshare for Agger). Bambauer (2013) notes the distinction between privacy and the capacity to maintain privacy (that is, the security required to maintain privacy). Equating the two means attention focuses on what is leaked (compounding the harm done). A greater attention to institutional failure to provide security places emphasis on those who should maintain it, not those who have fallen victim to failures to do so, or what is revealed.

2. From knowledge to information: control and informational capitalism

The dynamic between states, corporations, citizens and social movement actors creates a balance between competing undoings and assertions of privacy. All four sets of actors assert the public interest, but none truly represent it. The nearest thing to the defence of the public interest is the capacity of each to curb the power of the others. Technology neither determines nor is determined by social forces, as the two are never separate in the first place. The complex assemblage of surveillance-capable digital technologies as have arisen in recent decades has radically altered over time. Haggerty and Ericson's (2000) prescient overview of things at the start of the current millennium gives a sense of the transformation, both between the late 20th century and the year 2000, and in relation to what they wrote then and today's mobile networked world of devices and things.

For Giles Deleuze (1992), the information society is a new form of informer society, distinct from the knowledge society documented by Michel Foucault. Where, in Foucault's 'disciplinary society', power was exercised by experts within institutions governed by the disciplinary authority of professional expertise, today's 'society of control' steps outside such institutional spaces as the prison, hospital, school and barracks, and instead power circulates through networks in the form of information about fragmented 'dividuals'. The paradigmatic form of such information is a person's credit rating, in Deleuze's account. However, he also suggests new forms of self-administered 'electronic collars' will come increasingly to enable (and bar) movement and access to spaces, things, services and others in a networked society of controlling flows, monitors and modulations designed to track and shape dividuals into a new level of conformity. Signatures of professionals and numbers registering prisoners and pupils as subjects of treatment are replaced by codes for access and codes for being accessed. Private spaces are rendered private by the person accessing them using their access code. Facebook allows users to designate certain sharing spaces as private even in order that these are then to be used to share details about oneself with others (Bozarova, 2012). As such, gaining privacy requires identification. Where disciplinary society maintained surveillance through containment, the society of control maintains it by the monitoring of mobility. Debt and the desire to engage in constant retraining (for work and for correct consumption) require digital networks that both see us and engage us in wanting to see them: 'Man [sic.] is no longer man enclosed, but man in debt' (Deleuze, 1992, p 6). Debt and deficit fund and fuel control through information – to be informed and to be informed upon through the constant requirement to move on. Yet, this very negative account of the society of control stands in contrast to the sense (elsewhere in Deleuze's work; for example, see Deleuze and Guattari, 1984) that networks offer scope for fragmenting modulations of resistance in the rhizomic splitting and reforming that arises when agents of control seek to break up opposition. Hacker use of botnets can be seen as a manifestation of such rhizomic dispersal (Deseriis, 2016). The cat-and-mouse struggle between surveillance and privacy displays the capacity for new forms of surveillant access in information-based networks, but also new forms of concealment and evasion. Surveillance encourages developments in encryption and vice versa (David and Kirkhope, 2006). Each incites the further development of the other. That is the nature of cat-and-mouse modulations. Viewing privacy as a matter of trust, and not solely a set of legally defined rights, Waldman (2018) suggests that even when citizens do not know their 'rights', they are often right to distrust those they believe routinely betray their trust. Whether breach of 'trust' can become a strong foundation for legal redress is one thing. As a sociological explanation for anger, the sense of constant betrayal may be more significant in explaining resistance, even if other sociological factors are required to explain the 'privacy paradox' by which so much disquiet is so often contained within a society of control's 'business as usual'. When the police use Twitter to monitor the claims made by looters in a riot, the 'privacy paradox' moves to a whole new level (Cantijoch, 2013). Facial recognition technology may enable identification without any need to request documentation (Buckley and Hunter, 2011). However, when those accused of crimes counter with the use of surveillance footage to dispute accusations, the assumption that being 'caught on camera' undermines the legal presumption of innocence can be reversed (Hadjimatheou, 2016).

Shoshana Zuboff's 1988 book, *In the Age of the Smart Machine*, examined the relationship between automation and informating systems within the workplace. While the introduction of computers into the workplace did (and does) afford forms of automation designed to limit the control and awareness of workers, in relation to what it is they are working on and working to achieve, this affordance is not a 'logic' of technology as such, but rather the logic of a particular mode of capitalist accumulation. Presenting automation as a necessary method of increasing efficiency is, therefore, a particular kind of ideological strategy. Zuboff highlights the equally feasible affordance of computers in allowing what she calls forms of informating (in contrast to automating). Computers can afford a more informed workforce, and this can itself allow for a different kind of efficacy and efficiency that empowers workers, rather than a mode of automation designed to increase control by managers over workers. The themes developed in Zuboff's 1988 account of the workplace are developed in her later accounts (2015, 2019) of what she calls surveillance capitalism's new mode of accumulation (or what is referred to as 'data capitalism' by West, 2019). Where automation described a particular capitalist mode of accumulation in the realm of physical production and administration of services, today's surveillance capitalism seeks to turn attention into capital along a new frontier of capitalist development. Zuboff identifies four key dimensions to this development. The first dimension is the accumulation of data about people, the increasing capacity to access and integrate such data into cross-searchable form, and, finally, the ability thereby to analyse such data in ever more intense ways. This creates what Zuboff refers to as a new class of 'surveillance asset' or 'surveillance capital', paralleling earlier primitive accumulations based on primary extraction (agriculture and mineral extraction) and, later, industrial processing. The second dimension is monitoring and the creation of contracts that allow such monitoring to be expanded to the maximum extent (fighting specific legal battles along the way to defend general extensions that are undertaken and only then defended legally in retrospect). Contracts that allow generic access to user data as a default sets the power of surveillance capitalist enterprises (of which Google is Zuboff's exemplar) against individual users in a fashion that is so uneven as to render informed consent between equals dangerously close to obsolescence. Each individual stands alone and, therefore, largely powerless in the face of 'Big Other'. Fuchs (2011) notes the defining feature of Web 2.0 is the intransparency of data-capitalist enterprises relative to the transparency of such service users' data – and hence users themselves. Different ethical theories (Kantianism, Utilitarianism, Social Contract Theory and Virtue Theory; Hershel and Miori, 2017) emerged from different historical and political conditions. The question is whether today's conditions of 'big data' can be judged by earlier ethical frameworks. Whether new conditions require – or will generate – new ethical systems remains to be seen (Neff and Nafus, 2016). Andrejevic (2002; as well as Andrejevic and Gates, 2014) suggests the popularization of 'Big Brother' style reality television series in many countries, in which constant surveillance of 'real' people is packaged as entertainment, acts to normalize such increasing surveillance and its commodification in everyday life. The third dimension in the new frontier of capitalist development concerns personalization and communication moving from passive data collection and analysis to active behavioural modification as a mode of shaping user preferences, experiences and subsequent actions. Finally, the fourth dimension is continuous experimentation by surveillance capitalist enterprises, in which the scope and scale of further extensions to these three modes of informational capital accumulation is constantly being explored. The constant running up against specific objections, in the form of actions for privacy violation, becomes a routine business cost. Multi-million-dollar fines are modest costs when set against multi-billion dollar profits achieved by ever widening the scope for digital access. Defensive court actions become investments in the expansion of information harvesting and user management. Users manage themselves in the way they curate asynchronous exhibitions of themselves (as distinct from simply displaying themselves in synchronous performance). Whether, in digital contexts, users successfully manage what Goffman (1952) called the distinction between giving off (unintended) and giving (intended) information about their 'backstage' (real?) self (Hogan, 2010) varies. How far are users being incited by behavioural management programmes to breach their own intended limits? The capacity to curate must assume users have some agency in how they use services, even if they sometimes give away more than they wanted to – or may be encouraged to *want* to give away more than they might have otherwise wanted to have done, knew they had done, or might best be advised to have done.

3. Privacy, networked whistleblowers and the surveillance state of global networks

Whistleblowing precedes the digital age (Stranger, 2019). However, today's digital whistleblowers reveal the capacity of networks to breach the privacy of powerful actors who themselves often engage in digital infringement as well (Beckett and Ball, 2012; Coleman, 2014). The relationship between whistleblowing, traditional journalism and digital networks challenges established notions of authority, responsibility and the boundary between the public and private spheres (Benkler, 2011). WikiLeaks and digital whistleblowing more generally challenge the authority and capacity of states (and corporations) to keep secrets, even as revelations from whistleblowers have highlighted the increased capacity of state and corporate actors to breach the privacy of others (Brevini, 2017). The rise (and perhaps also the fall) of WikiLeaks highlights the nature of digital whistleblowing and its relationship to digital journalism. WikiLeaks' name suggests a space for user-generated content, and the site does act to provide a secure space for whistleblowers to release sensitive material. However, the site has also become increasingly engaged in the selection and curation of content, which takes it in the direction of something akin to journalism, with the issues of liability for content that this implies (Lindgren and Lundstrom, 2011). Founded in 2006, WikiLeaks came to prominence with the publication of leaked documents highlighting corruption in Kenya. The site gained even more notoriety after releasing an edited version of leaked US-military-helicopter camera and audio footage showing the shooting of a group of civilians in Iraq in 2007 (titled 'collateral murder'; see Christiansen, 2014). This release came in 2010, as part of a large leak of classified documents made by the then US army private Chelsea (at the time, Bradley) Manning (Mader, 2012), and marked a separation with a number of respected newspapers over the degree of redaction/revelation applied to leaked content (Stöcker, 2011). Manning's leaked material became known as the 'Iraq War Logs' (see Beckett and Ball, 2012). Manning's court martial in 2013 (Rotte and Steinmetz, 2013), but release in 2017 (Maxwell, 2018), can be contrasted with the ongoing detention of WikiLeaks' founder Julian Assange (DW Documentary, 2020). Later revelations of US presidential candidate Hillary Clinton's private emails in 2016 led to allegations that WikiLeaks had, knowingly or otherwise, acted as a political pawn on behalf of those who had hacked Clinton's emails and wished to see the election of Donald Trump.

In 2013, former US National Security Agency (NSA) contractor Edward Snowden revealed the scale of global (and national) electronic surveillance conduct and capability (Greenwald, 2014; Snowden, 2019). In particular, Snowden revealed the scale of cooperation (required by law but also kept secret by law) that occurs on a number of different fronts: between US digital service providers and the NSA; between the NSA and intelligence agencies in the UK, Canada, Australia and New Zealand (the five eyes), especially in surveilling each other's citizens (getting around some residual legal limits on mass data collection and analysis on each state's own population); and in surveilling citizens, politicians and other organizations within 'allied' countries, as well as within states actually/supposedly hostile (see conflict between the European Union and the UK/US over the Eshelon surveillance programme: www.europarl. europa.eu/doceo/document/A-5-2001-0264_EN.html). The very idea that civilian communications, *en bloc* and *in advance*, should be treated as 'signal intelligence', something previously reserved for specifically military communications, is a radical and highly consequential shift (Andregg, 2016). The scale of this activity, in tapping cables (requiring collection by commercial service providers) and in processing such material, went far beyond what was previously thought either possible or actual. Meta-data (who connected with who/what, and when) for both internet and telecoms, and then more specific data (what was seen/said), was now being accessed and processed on a previously unimaginable scale, in both the level of intrusion in particular populations, but also in global reach. Bauman et al (2014) suggest Snowden's revelations highlight the need to go beyond a simple idea of the extension of the Big Brother 'authoritarian' state, to see, rather, the reality of a new transnational, digitally networked and global surveillance 'state' that exists beyond nation, democracy and accountability. Lyon (2014) adds that Snowden also revealed the scale of 'big-data' expansion and integration in both collection and analysis. Yet, these authors also highlight how Snowden was able to demonstrate what he did via the very networks that have afforded the surveillance creep he himself revealed. The reaction, worldwide and in the countries most heavily implicated by the revelations (the US and UK), show that resistance is possible and that whistleblowers can make a difference. Gabriella Coleman (2014 and 2019) adds to this by highlighting how hacker groups like Anonymous, and other hacker activists, have developed and distributed new forms of encryption in response to the kinds of revelations made by whistleblowers like Snowden (who himself helped develop the encrypted messaging app Signal – other hacker involved systems – like SecureDrop – are available).

Online dissemination of revelations by such organizations as #MeToo and Black Lives Matter – whether in the form of whistleblowing testimony by victims of sexual abuse by employers such as via #MeToo or in the posting of filmed footage of police violence against African Americans – adds a different dimension to the digital scope for revelation and counter-surveillance. Whistleblowers have also posted materials online in relation to the COVID-19 pandemic, such as those made by the Chinese doctor Wi Wenliang. Disputes over when it is right to betray one's employer's privacy (Tavani and Grodzinsky, 2014) are particularly complex when the 'betrayal' in breaching an employer's privacy is in the context of employers whose job is to breach the privacy of others (such as in the Snowden case). Framing such actions as civil disobedience or criminality (Scheuerman, 2014), or as between civil disobedience and vigilantism (Delmas, 2015 and 2017), becomes a key method of legitimating or discrediting such actions. Such framings hinge on divergent conceptions of 'the public interest' (Qin, 2015). Drawing on the work of Niklas Luhmann (discussed in Section 1), Thomas Olesen (2019) suggests that the digital networks challenge system and subsystem boundaries. As such, more and more communication (flowing across such boundaries) will likely come to constitute what is called 'whistleblowing'. Breaching (communication across institutional boundaries) does not even have to be intentional to be significant, but the drama that unfolds from field transgression, over who has the right to speak, becomes ever more prominent in an age of horizontal communication (Olesen, 2018). Different states have different laws defining data breaches, and concerning whether/when such digital release may or may not constitute the legally protected act of 'whistleblowing'. However, in line with Luhmann's conception of breaching – because most people do not know what those laws actually are, and because of the prominence of the United States within global media culture (online and otherwise) – it is increasingly difficult for states to deny that they support the rights of whistleblowers, even if their formal laws do not technically do so. Access (both in gaining compromising information and in making it available to others) displays a significant level of symmetry. Powerful state and private agencies enact surveillance on populations/users, even as such actors are themselves subject to whistleblowers who reveal even that which their paymasters had paid them to access/conceal in the first place.

4. Hackers and hacktivists

The supposed evolution of the idea of hacking – from technical play to malicious harm, to politically motivated rebel hacktivism and then to state-sponsored information warriors – is not a logical progression of any 'spirit of informationalism', where information 'longs' to be free. One person's play is another person's vandalism and yet another person's revelation. The colour of a hacker's hat (black, white or grey) is in the eye of the beholder, not an objective classification. The term 'hacker' means many things, as does the act of performing a 'hack'. To hack, as in to chop something up, becomes a metaphor in relation to computing devices and networks when some kind of modification or repurposing takes place. The distinction between hacking and cracking, between modifying and breaking (in or up), is again open to interpretation – as is the question of authorization, when the question of who might give such an authorization is as much in question as the valuation of that which is being altered and of the alteration.

Tim Jordan (2017) offers a 'genealogical' approach to mapping this diversity, following Michel Foucault's call to map the way histories are constructed as tautologies of arrival (in the present) rather than as neutral documentations of historical change. For Foucault (Jordan argues), 'genealogy' avoids singular origin stories, explores the novelty of the seemingly natural and seeks to unearth the forgotten alternative paths in what precedes a present that conceals its own contingency by omission (see also William, 2011 on what history and historians tend to forget). Applying these principles, Jordan seeks to unearth the contradictory history of what is now called 'hacking', in four parts (though these 'ages' overlap and return, so are not 'developmental'). First, what starts with the repurposing of hardware, and then develops into what is now understood as programming (the pre-history of hacking), saw a certain gendered (masculine) culture of control and play emerge through peer education, do-it-yourself showing off and the development of networks of such learning, showing and valuing. This new cultural and technical (networked) space afforded, second, 'the golden age of (C) racking' (Jordan, 2017, p 534), with its ideal-typical access geeks accessing and modifying other people's computer systems simply because they could. Third, this space then divided into criminally and politically motivated hacking on the instrumental side, and between free and open source forms of programming culture on the intrinsic/creativity-for-its-own-sake side of so-called hacking. Finally, there is an incorporation of hacking into state and corporate practices – whether in the form of informational warfare or in the development of ever more sophisticated behavioural management of service users. What is significant in this genealogy is that each moment is in constant reappearance, and the meaning of 'hacker' is all these things at once, such that one person's conception of what hacking is can always be countered with manifest examples of its opposite. The tensions manifest within politically motivated hacking ('hacktivism') – between mass-action hacktivism (designed to crash or disrupt), digitally correct hacktivism (promoting informational freedom, itself torn between the desire to reveal what elites encrypt while also giving non-elites effective encryption to avoid state and corporate surveillance) and dot.Communist hackers (keen to make access to copyrighted content free) – illustrate this diversity within just one dimension of hacking's genealogy (Jordan and Taylor, 2004; Jordan, 2008). The tension between masculine control through programming and forms of politically progressive hacktivism can be seen in Nagle's (2017) account of trolling within the online gamer world (as discussed in Chapter 3). Tensions between anti-state- and state-based hacking were also noted in Chapter 2. The claim for a 'hacker ethic' (Himanen, 2001; Kirkpatrick, 2002), founded in information's desire to be free, must be set against the multiple ways in which the hacker is always already embedded in other logics (whether these be gendered, commercial, criminal or security-related). Yet, members of hacker networks themselves engage in discursive stabilization, the ongoing attempt to stabilize the meaning of what they do, who is to be included and who is shut out (Lindgren and Lundstrom, 2011). It is not just an external gaze that seeks to order and classify, and, in many ways, to simplify and historically delete.

In contrast to Jordan's genealogical approach, Kevin Steinmetz (2016) maps the history and practice of hacking within a political economy perspective drawn from a Marxian approach to work, culture and power. Where Jordan's approach is to study the construction of hackers, Steinmetz seeks to study the practice of hacking from within, and, in particular, the 'craft' and 'crafty' nature of hacking, as understood by those engaged in it. While distinct in its approach, Steinmetz also draws out a similar set of contradictions within what hacking is thought to be, along lines of gender, class and power, not entirely dissimilar to Jordan. Where Jordan offers a four-part history, though, Steinmetz suggests a six-part scheme (again, history and classification blend as timescales overlap and elements repeat): hardware modification; gamers and software programmers (free, open source and commercial); phone phreakers (action at a distance); bulletin boards and networks; cryptographic politics (in defence of privacy); and the rise of techno-transgression/hacktivism more widely (in the face of state/corporate colonization of programming). In Steinmetz's view, what marks out hacking is its particular craft/crafty performance, rather than any particular conception of its good or malicious outcome. Hacking is many things. How hacking's framing shifted from geek to criminal says more about state, corporate and mainstream media hostility, towards decentralized communication and any challenge to centralized ownership of property and political power (Nissenbaum, 2004; and see also Dizon, 2019), than about how hackers see themselves or what motivates them. Framing hackers as terrorists is another common representation (Vegh, 2002). Yet, being labelled (and often prosecuted) as a 'hacker' by authorities becomes a positive identity for many so labelled (Turgeman-Goldschmidt, 2008). In addition, hacktivists have also tended to reify their own practices as flowing from certain technological logics – in particular, the idea that information wants to be free (Söderberg, 2013). The actions of the NSA and its subcontractors, and the actions of Facebook and its subcontractors and corporate data users, can reasonably be described as hacking. Yet, the way hacking is framed in legal and political debates focuses on the threats posed by challengers rather than the whistleblowing potential (and the encryption-enhancing potential) of anti-establishment-oriented hackers in relation to dominant actors who themselves hack networks to their own advantage.

Luke Goode (2015) identifies the contradictions within the Anonymous hacktivist network, combining elements of the nihilistic (and often sexist) pranksterism of earlier 4Chan networks, like LulzSec (Arthur, 2013), with more egalitarian goals, and promoting privacy at the same time as undoing the privacy of states and corporate actors (Stryker, 2012). Moreover, a libertarian ethos of demanding the right to be left alone was combined with, but was also in tension with, other collective goals. While WikiLeaks was strongly identified with one person, Julian Assange, Anonymous actively resisted identification with any one leader or personality (Coleman, 2011). However, social network analysis techniques used to map connections between nodes within hacker communities like Anonymous (see Decary-Hetu and Dupont, 2012) were also used by law enforcement actors to identify key nodes in the Anonymous network (see Chapter 2). The valuation of key actors within software piracy networks can also be used to track and identify such contributors by authorities – in effect, reverse hacking the hackers' own hacking networks (Decary-Hetu, Morselli and Leman-Langlois, 2012). Yet, hacking can emulate some perfectly legal forms of protest (Hampson, 2012), just as some kinds of copying are protected in some jurisdictions. Adam Klein (2015) found that while most mainstream media coverage of Anonymous was hostile and focused on its criminal aspects (in terms of unauthorized access, damage and disruption), the great majority of the actions being reported on in such a context were in defence of free expression, which is itself something that most states and most media outlets claim to support. The 'trickster' characterization of the hacker is riddled with contradictions, in terms of both hacker self-identification and wider societal framings (Nikitina, 2012).

5. Hacking, cyber warfare and spying online

It was once falsely imagined that the lone, malicious, hacking 'devil' drove a Lada, and now we are told the St. Petersburg Internet Research Agency has weaponized our folk-devil into a real political force. David Wall (2012) argues that, in the first decade of the 2000s, there was a common stereotype about the hacker, as being a highly educated, but economically insecure, young Russian man. The idea of 'the wild East' threatening stable digital interactions in the West was at variance with data suggesting that most hacking was 'local', such that where there were high levels of internet use, so too would there be high levels of hacking. After 2010, however, Wall argues, the revelations of WikiLeaks (and, after Wall wrote, of Edward Snowden) put the spotlight back on to Western governments, and Western hacker groups like Anonymous and LulzSec, as the most significant source of hacker attacks, at least at the political level (fraud and blackmail-based hacking will be addressed in Chapter 8). It is interesting, then, to note that, after the 2016 US presidential election, attention has shifted back to the image of the super-powerful Russian hacker as the paradigmatic politically motivated online actor. The suggestion is made that the hacker has been weaponized by states like Russia (along with China and North Korea), and that the hacktivist has now been replaced by the patriot hacker/cyber warrior. Yet, fundamentally, Wall's suggestion remains true: the stereotypical hacker driving a Lada (that is, a Russian) is a myth. While there may have been recruitment by states of previously autonomous hackers to now work for the government, this is true for all states, and all the more so for those in the West who suggest they are now threatened by other states doing what they themselves are also doing.

Determann and Guttenberg (2014) argue that spying remains legal in countries employing particular spies, even as it is illegal in those countries where such spies are sent. All states permit state agencies to collect intelligence on other states; and all states make it an offence for other state actors to spy on them. International law does not make it a crime for individual states to make their own spies legal, even as international law does not prohibit states from criminalizing the spying of foreign state actors upon them. Treaties between states (multi-lateral and bi-lateral) may regulate relations between states, but treaties are not the same as domestic law. Domestic law in relation to the treatment of a country's citizens is different from treaty obligations that may temper, but do not determine, how states interact with other states (or with the citizens of other states). All states treat foreign spies as enemies and their own spies as heroes. Digital networks mean that intelligence can be gathered remotely. The distinction between domestic law and foreign actions, both military- and intelligence-based, and, more particularly, between the police and the security services (often considered a fundamental distinction in liberal democratic societies), breaks down if local and remote actions and the surveillance/regulation of both begin to blend into one another. Determann and Guttenberg suggest the wall between policing and security was significantly dismantled in the years after the 11 September 2001 (9/11) terrorist attacks. Police actions against drug cartels draw heavily upon military intelligence. Action against human traffickers, likewise, draws together security and intelligence service information with physical law enforcement actions in operations across borders (Gerry, Muraszkiewicz and Vavoula, 2016). This breaking down of the law/war agency/rules is true in the United States but also of other states in an increasingly, digitally and physically, connected world. States, while publicly outraged by Edward Snowden's 2013 revelations about the US NSA, were themselves already engaged in much the same kinds of practice (even if on a much smaller budgetary and geographical scale to that undertaken by the United States). Determann and Guttenberg (2014, p 889) observe that 'we still live in a dangerous world'. If one state were to reign in its intelligence capability, this is not likely to deter others, but rather perhaps to encourage other states to take advantage of any vacuum created. However, there is an irony in the rapid expansion of hacking capability, and the absorbing of many computer programmers, who might have otherwise engaged in non-state-backed hacking, into state cyber-warfare security agencies. Substantial expansion of intelligence gathering by states post-9/11 went hand in hand with an increased privatization of such work (to subcontractors). As Bauman et al (2014) noted (in Section 3), privatization of intelligence, along with networked and global forms of such data gathering, creates a global intelligence assemblage more powerful than any traditional 'authoritarian' state; but such assemblages are also weaker in that they leak like sieves – with Edward Snowden being only the most visible example. All these hackers working for the government, or for subcontractors to the government, become potential whistleblowers. As professional hackers, they may moonlight as hackers of the very agencies that pay them to be hackers elsewhere. Bauman et al go so far as to suggest that, as a global intelligence 'guild', hackers within intelligence institutions have no more loyalty to their employers than state institutions have to the citizens they claim to work for (with nation and state increasingly disentangled in global networks of power, money and information). The battles in the 1990s over encryption between states and 'digitally correct hacktivists' have not disappeared (Moore and Rid, 2016). It is now simply clearer that states back then were employing 'digitally incorrect hackers', as they do now. Now, states simply pay more of them, even if such 'digitally incorrect' patriot hackers are more likely to be in precarious (subcontracted) employment today. Hackers worked on both sides, swapped sides and occupied all points in between; they still do, but may only now be a little more likely to say so.

In conclusion

Digital surveillance has increased scope to access the computers of others, and, hence, to invade forms of privacy that are themselves only the recent product of digital networks themselves. States (Bauman et al, 2014), corporations (Zuboff, 2015 and 2019) and other individuals and groups (Jordan, 2008, 2017; Steinmetz, 2016) have extended capacity to access the lives of others, but are themselves also subject to such invasions of security and privacy. While hacking (in its widest sense – as practiced by states, corporations and others) has increased access, hackers have also been at the forefront of developing and disseminating forms of encryption that allow new levels of concealment. The contest between surveillance and encryption, just as with any technological affordance, can cut in both directions, and outcomes are always contingent social processes, not technologically determined effects. Fuchs' intransparency of data-capitalist corporations in relation to the transparency of users is contingent and can be resisted (although most people do not even try). State efforts to conceal their own surveillance has been undone by whistleblowers. Digital action at a distance does mean today's spies can evade the laws of countries they spy upon more easily than ever before, even as whistleblowers like Edward Snowden can evade prosecution by more physical means of distancing (travelling to another country). Chelsea Manning was not so lucky and did not evade prison, as her concealment was not strong. However, she was pardoned after three years, which is evasion of a kind. Julian Assange remains in a UK prison, after spending many years in London's Ecuadorian embassy. The United States has yet to successfully extradite him, which is evasion of a sort, although incarceration in Her Majesty's Prison Belmarsh is not freedom. Despite supposed legal safeguards, the cost of whistleblowing can be very high (Kenny and Fotaki, 2021). Whether Facebook and Google's behavioural management algorithms constitute 'hacks' is one thing; how far they successfully incite users to buy certain products or vote in certain ways is another. While the potential for symmetrical access and evasion exists, in practice, states and corporations have extended the affordances of digital technology in their own interests, far beyond what citizens, consumers and activists have been able to counteract. However, competition between states and corporations, between states and between corporations, creates spaces for contradiction and disruption to arise which continue to enable challenges and resistances to be enacted.

7

Fake News, Echo Chambers and Citizen Journalism

Key questions

1. Are citizens today more informed or more misinformed than those in pre-networked times?
2. Does free online content offer an alternative to corporate monopoly sameness, or has the rise of free alternatives undermined creativity and fuelled the commercial focus on safe sameness?
3. Has the internet opened up scope to explore difference or encouraged increasingly detached silos?
4. Are censorship and editorial control the answer to fake news or the cause of it? Are traditional print and broadcast media more or less biased than online content is diverse?
5. Do digital networks cause increased social polarization or is it simply error to 'blame the media' for social divisions?

Links to affordances

Digital media increase the scope for citizens to access alternative news content, some of which is misinformation. It is not always clear to those accessing such content, or who are accessed by those seeking to deliver such content to them, who is packaging that information. Concealment remains, although this is true of mainstream media as well. Moreover, a part of online discussion, whistleblowing and alternative media production also serves to expose such actions. However, how far, once identified as misinformation, such content can be blocked is contested. Some content gets through (evading control), but blocking by search engines is prevalent – raising its own concerns over manipulation. Mainstream media rely in large part on such sources too. Claims that audiences are incited easily by such content are exaggerated. While existing tensions can be stoked, such incitement cannot take place in a vacuum, and state/corporate media are no less responsible for seeking to incite/manipulate audiences – with equally mixed results.

Synopsis

The internet has created the potential for 'everyone' to publish their own news and other content, and for 'everyone' to access it. Whether such 'citizen journalists' are treated as 'journalists' in law varies by jurisdiction. On the one hand, citizen journalism has undone censorship, while, on the other hand, it has led to fears of a loss of standards. The same arguments, between those that value editorial control and those that decry it, take place in relation to academic publishing and to trade writing (fiction and trade non-fiction). To what extent is cyber culture creating deeper social silos than existed before? Does new media increase or decrease the volume and susceptibility of viewers to so-called fake news? It is not the case that people today are narrower in their choice of news. Once, people read a newspaper that reflected their preconceptions, while the state or advertisers largely dictated radio and television's ideological content. Today, algorithms play a role in reproducing such an editorial nexus. Censorship and bias are not new, nor are they likely worse than ever before. If society is more divided today, new media reflects this division, even if it may afford exaggerations of such divisions. While media may reinforce either cohesion or division, such as may exist outside it, the idea that 'new' media intrinsically polarizes where old media unified is false. What divisions in society exist need to be addressed, even if it is always easier to shoot the messenger.

Chapter sections

1. Within today's new-media networks (distributed digital networks of user-generated content), an increased amount of information is no guarantee of truth, and even less of wisdom; but misinformation (and/or disinformation) online today is no greater than was afforded by traditional print and broadcast media in past times and at present. Editorially managed media outlets have also undertaken their share of criminal hacking (Freedman, 2012; Wring, 2012) and bias. One person's professional journalistic editing is another person's censor.

2. Global network capitalism displays divergent processes of increasing global media concentration of ownership within film, books, music, software and so on – even while digital networks also enable distributed copyright infringement of such intellectual property by means of online sharing. Free sharing may cause commercial narrowing via churnalism and reliance on establishment sources; but the example of scientific research illustrates

that free sharing can foster wider collaboration and better knowledge (Castells, 2009; David, 2017a).

3. It is possible to find your tribe online and then to bury yourself within that community (Balmas, 2014). While potentially liberating for those that feel marginalized in real-world situations, such silos may be socially divisive and inciteful of hostility; yet, yesterday's daily newspapers were far narrower, and readers more loyal to their blinkered sources, than are today's 'disloyal fake news audience' (Nelson and Teneja, 2018).

4. While imperfect and requiring a willingness to critically engage with professional journalistic values and routines (Harcup and O'Neill, 2017), new media does show that more information is better than censorship (Garton Ash, 2016).

5. Blaming the media oversimplifies issues. Blaming new media is likewise. Fake news can fuel emotional economy blame and reduce scope for dialogue, increasing social and political hostilities (Bakir and McStay, 2018). However, state and commercial media are guilty of such ideological stoking too. Neither Vladimir Putin nor Rupert Murdoch won the 2016 US presidential election for Donald Trump (Hillary Clinton actually got more votes). Fake-news websites exerted influence within a limited and predisposed, partisan audience (Vargo et al, 2018).

1. Where is the knowledge lost in information?

The increased amount of information in circulation in an age of user-generated new-media platforms that circulate and promote content on the basis of likely re-circulation (and/or concealed payment) rather than any journalistic/editorial principles is no guarantee of truth, or of the wisdom to understand it; but levels of misinformation (and/or disinformation) online are no greater than in traditional print and broadcast media (where editorial control may be no more or less open to abuse). It should be recalled that new-media platforms that allow all users a voice are themselves businesses with an interest in promoting sensation (which encourages engagement and, hence, more advertising revenues). Whether editorially controlled 'old' media or algorithmically managed 'new' media offer more truth is therefore complex. As Natalie Fenton (2010) argues, we should not assume in advance that either techno-optimists or techno-pessimists are right when it comes to claims about new media and the future of news. Over a decade ago, Luke Goode (2009) called for attention to be paid to the following processes: the tension between peer-to-peer forms of horizontal communication and more hierarchical conceptions of knowledge within traditional media structures of authority; the productive interplay between the two in creating a more informed society; and the creation of new forms of agenda-setting power within such interactions. Over a decade on since Goode's suggestion of this research agenda, the issues raised have only become more pressing, and the tensions identified have only become more obvious.

The term 'fake news' is routinely associated with new media. However, the engagement of many editorially managed media outlets in criminal hacking and bias make them no better than online alternatives (even as some editorially managed traditional media outlets can reasonably claim not to sink to such levels). In 2011, it was revealed (by an investigative journalist working for the UK's *Guardian* newspaper) that another UK-based newspaper (the *News of the World*) had (in 2002) paid a freelance investigator to hack into the mobile phone voicemail account of a murdered child, creating the impression that the child was still alive. The newspaper sought to gain some advantage from the information revealed in the messages, but seriously damaged the police investigation into what only later became a murder inquiry, and caused significant false hope and then distress to the family of the then missing, and presumed still alive, child (Millie Dowler). These revelations led to the closure of the *News of the World*, and the creation of a public inquiry into press intrusion (the Levison Inquiry, which was drawn out, watered down and then shelved). The Inquiry found the practice of phone hacking and illegal breaches of privacy were common practice in a number of British mainstream newspapers – and not just the *News of the World*. Later investigation saw the criminal conviction of the former editor of the *News of the World* (Andy Coulson) for involvement in a separate instance of authorizing phone hacking. He had earlier (2007) resigned his position as editor in relation to a previous incident for which he denied knowledge, and was (at the time of the 2011 phone-hacking revelations) the director of communications for the British prime minister David Cameron (a position Coulson resigned from as the storm erupted over his previous activities). Coulson was sentenced to 18 months in prison, but served only around 18 weeks (even as his editor escaped criminal liability entirely due to the English legal test for corporate liability). As Wring (2012) and Freedman (2012) suggest, mainstream media editors, the media corporations that employ them and the politicians who seek their favour exercise a controlling and corrupting influence on public life.

Illegal invasion of privacy, political bias and the distortion of news are things new media are routinely accused of having brought to 21st-century journalism, but such practices are not new. Moreover, the United Kingdom is not unusual. Commercial interests dominate content and perspective in US news, even as state controls in countries like China and Russia mean that mainstream editorial control acts as a system of political censorship rather than in maintaining standards of objectivity, impartiality and/or balance. It may be argued that public service broadcasting in the UK (in the form of the BBC) is less biased and more balanced than commercial rivals, but, even here, the pressure placed on such broadcasters to avoid direct confrontation with governments, who (directly or indirectly) determine their management, funding and charter/mandate, means they are still often accused of bias in favour of economic and political elites.

In recent years, new-media platforms have been accused of fuelling conflict and hatred. Twitter storms in India have fuelled intercommunal acts of violence that have led to many deaths. Narrow casting of newsfeeds is said to have created filter bubbles and silos that increase division and hostility between in-group and out-group members (alongside the decline of common ground shared through more widely shared/broadcast media and 'common culture'). Targeted political messaging has been accused of polarizing political opinion. The history of print and broadcast media (what are now referred to as old media in distinction from network-distributed digital 'new' media) can hardly be said to be untainted by all these accusations. In 1753, the Jewish Naturalisation Act gave equal rights to Jews living in the United Kingdom, but the Act was repealed a year later after riots stoked by a virulently anti-Semitic, early national press. Patriotic newspapers stoked national sentiment on all sides in the run up to and conduct of the First World War (1914–18), which saw the deaths of tens of millions of people. Communist and National Socialist exploitation of newspapers, film and radio, likewise, contributed to the murder of even greater numbers in the middle years of the 20th century. State radio broadcasts in Rwanda in 1994 incited civilian militia groups to massacre around a million fellow Rwandans (labelled as 'other' using 'tribal' identities solidified during European colonial rule in earlier generations), and sparked a war in central Africa that killed millions more. Such horrific associations between mainstream media and state violence, censorship and ideological bias have led many to place faith in new kinds of online 'citizen journalism' to redress the balance. It is not perhaps surprising that defenders of mainstream media decry such amateur reporters as unprofessional and ill-informed, but is this just 'sour grapes' or do professional journalists have a point? Are we now living in an age of 'fake news' and 'filter bubbles'?

The last 50 years has seen the rise of unmediated media. Abraham Zapruder's 14 seconds of film footage captured the assassination of John F. Kennedy in 1963. The footage was bought by *Time Life* magazine and kept from public view for over a decade. In 1991, the beating of African American Rodney King by Los Angeles police officers was filmed by local resident George Halliday using an early digital camcorder. Halliday, like Zapruder, sought to sell the footage, but when unsuccessful, released the digital material directly online (again in its infancy, at least for civilian use), fuelling widespread protests. In December 2004, the Asian (Indian Ocean) Tsunami saw hundreds of thousands of people killed and millions made homeless. Many of those caught up in the event recorded their own experience, and sometimes even their own deaths, on mobile phones with digital cameras directly linked to the internet. This event saw direct experience and unmediated access to an audience surpass the capacity of professional journalists in timing, access and insight into the tragedy. Later extensions of this unfolding of 'unmediated media' saw rebels, who overthrew Colonel Gadhafi in Libya in 2011, record and broadcast his killing on mobile phones; saw the spate of beheading videos discussed in Chapter 2; and saw the murder, live on television in 2015, of US news presenter Alison Parker and her cameraman Alan Ward, by a disgruntled former colleague who recorded and broadcast his crime live on the internet. To date, the most prominent case of unmediated media was the filming on mobile phones, and the immediate circulation of this footage online, of the murder of George Floyd by a police officer in 2020. Where critics of citizen journalism claim direct reportage of experience by non-professionals lacks context, supporters point to the immediacy of facts, and the ability of such facts to be revealed in ways that earlier generations of managed/mediated (that is, editorially controlled) media left concealed. Shoemaker and Reese (2014) argue that a series of mediating contexts (institutions, organizations, routines and systems) remain in the structuring of media effects upon audiences, such that there can never be any such thing as a fully unmediated media. However, potential to access a diversity of content has increased, even if we should therefore be all the more attentive to the ways in which what we do access (by our own choice or by the influence of others) remains partial.

Citizen journalism covers a wide range (Wall, 2015), including hyper-local news gathering and dissemination, the fusion of local actors and audiences via global platforms (like blogs and Twitter), globally minded activists engaged with highlighting local manifestations of issues, and the various uses and blending/decrying made of local sources by mainstream media 'professionals' (who will often criticize but use locally sourced content, if they can). What Stuart Allen (2007), after Manuel Castells, calls 'mass self-communication' does have the power to give new insight to audiences and a voice to those caught up in events designated as 'news'. Allen gives the example of online citizen journalism by those caught up in the 2005 London bomb attacks. Such citizen reporting reframed the story to one of local resilience and community. However, it should not be assumed that togetherness is the inevitable outcome of local coverage. Yet, in looking at local 'crisis' reporting by those in the midst of events, Allen and Thorsen (2009) suggest that, overall, citizen journalism on every continent is adding to the creation of a better informed world (see also Allen's 2013 discussion of 'citizen witnessing' in this regard). Tewksbury and Rittenburg (2012) suggest a combination of benefits and harms, with the blending of old and new sources affording greater insight, but with the tendency of audiences to fragment in their consumption of news along lines of existing cultural and political affiliations, the generation of increasingly local news for presumed audiences of like-minded people diminishes the demand for and the capacity to offer balance. Carpenter (2010) found that, contrary to the concern that citizen journalism would tend to be more insular than mainstream media in its use of sources, the reverse was the case. Content diversity within citizen journalism, combined with its additional contribution to news generation (relative to just having older media), means, Carpenter argues, that audiences are better informed today than in the past (although the past may not be a very high benchmark). Online alternatives to traditional media challenge established models of funding (in particular, directing advertising revenues from television and newspapers to online platforms). This makes the production of traditional news harder to finance – producing an increased reliance on citizen-generated content, which may be both positive (in terms of diversity of voices) and negative (in terms of the reliance on sources with vested interests). In either case, the future to be adapted to is already here (Anderson, Bell and Shirky, 2015). Neither the past nor the present are perfect, but neither is the present fixed or monolithic. Nguyen and Scifo (2018) distinguish citizen witnessing, political advocacy forms of citizen journalism and expert voice modes of non-traditional journalism. The benefits of unfiltered (uncensored) news need to be set against the dangers of 'release, then filter' forms of unwarranted and unchecked claim making, and the danger of particular communities having access to/acceptance of only pre-filtered, and subsequently polarized, belief.

2. Concentration and distribution

Global distribution (access) offers greater profit to increasingly integrated transnational firms, even as digital distribution also affords evasion and concealment when it comes to paying tax on those profits. The rise of global network capitalism with the end of the Cold War saw the divergent processes of an increasingly concentrated global media system within film, books, music, software and so on – even while new digital networks also afforded unprecedented copyright infringement of that content in the form of online sharing. Some argue that free sharing has caused commercial players to become increasingly narrow in their focus on churnalism and establishment sources; but the example of scientific research illustrates that free sharing is the best way to better knowledge.

Global publishing is now dominated by the English language medium, and, since the 1990s, increasingly by English language media corporations. Until 1988, the United States had not signed the Berne Treaty on copyright, and did not acknowledge copyright on any work not physically produced within its borders. In 1988, the United States did sign up to Berne. After the 1994/5 TRIPS Agreement (Trade-Related Aspects of Intellectual Property), under pressure from the United States, all countries (under threat of trade sanctions) were pressured to sign into domestic legislation the protection of foreign copyright holders. Since then, the scale of global media concentration has intensified (Castells, 2009). This involves cross-national integration: so, for example, most books sold in the UK are published by companies owned by the same four top publishing houses that sell most books in the United States (Thompson, 2005 and 2012), and these companies also dominate markets worldwide (even as non-English language media companies have in some cases shown strong resistance to such homogenization). Newspaper and television channel ownership is increasingly concentrated in this manner. Concentration also involves horizontal integration (with film studios, record companies, publishing houses and television/newspaper, as well as ISPs, being increasingly integrated within cross-media, multi-national corporations) and vertical integration (such as where record companies own the music-publishing houses that hold performance rights on songs, or where holding companies own film studios, distributers and cinema chains). The same ISPs that offer users access to a greater range of media content are also themselves increasingly integrated within concentrated cross-media ownership; and a small number of social media platforms (Facebook and Google for the most part) scoop up a greater and greater share of global advertising revenues at the expense of older and more diverse national providers of media (and specifically news) content.

News is one dimension of an overall media landscape. As with broadcast and print news, so with trade publishing. In recent decades, there has been a global concentration of book sales around what publishers refer to as 'big books' and, of these, those sold by the largest global publishing houses (and their raft of subsidiary imprints – which create the impression of variety). Thompson estimates that in the first decade of the 2000s in the UK, the number of titles selling more than 10,000 copies a year fell from around 600 to 450, with almost all of these being TV tie-ins and/or 'repeaters' by famous, formulaic, fiction (genre) authors. The number of copies sold of each such 'big book' has increased as the overall number of such 'big books' has declined, and the number of works selling between 5,000 and 10,000 copies has also declined. Anderson (2009) notes a similar concentrating top end, and thinning middle, in the United States. However, Anderson also notes the rise of 'the long tail', the basis for the Amazon business model. While bookshops make most of their money from high volume sales of a very narrow range of 'pot-boilers' (TV cookbooks and genre fiction) and repeaters by established authors, Amazon makes more money from the fewer sales gained on each of the many millions of works it can offer via its distributed storage and delivery model. While Thompson estimates that only 1 per cent of UK authors sell more than 10,000 books a year, and Anderson estimates that only 2 per cent of authors in the US sell more than 5,000 copies (neither of which would secure them an income sufficient to live on), the remaining 98–99 per cent live by other means. The long tail gives these (non-formulaic, non-TV-tie-in, non-pot-boiler) writers some audience at least, even if (as Jessica Silbey, 2015 argues) the vast majority of such authors make a living by teaching, journalism, ghost writing, editing, giving readings, performance and other non-authorial activities in the support of their creativity. Shakespeare wrote a century before copyright was first created (in 1709). Charles Dickens made more money doing live performances in the United States (where his works circulated without copyright and, hence, created a larger audience) than he made from selling books at home. Mozart and Beethoven had no copyright on their works in Austria. English contemporaries did in their home country, but who remembers them? As Castells points out, global networks have afforded an unprecedented level of concentration, even as networks have also afforded the long tail that Anderson suggests can offer some degree of resistance to such concentration. The capacity of fans to develop fan fiction has increased, while clumsy attempts by record companies, publishers and film studies to clamp down on this can have a harmful impact on the very commercial products such companies are seeking to promote (Potts, 2012; Liebler, 2015).

The field of academic publishing offers another parallel with that of trade publishing (and journalism). Since the 1960s, more and more academic journals have been bought by commercial publishers. Prices have been raised year on year (as the core journals serving particular academic fields are not ones that universities can readily cancel from their library subscription lists). Thompson (2005) calculated that an average increase of 13 per cent per year over four decades has produced a many thousand per cent price increase (which has carried on apace in the time since he wrote). As digital networks mean journal articles can be circulated freely, their price continues to escalate. Authors are not paid (and sometimes even pay to publish), and peer review is also undertaken without payment. The free circulation of ideas (what Merton, 1942 called 'academic communism'), by which status is accorded in academia, sees careers developed through freely giving away ideas and having them cited by others – with not being cited as the mark of career failure; Charles Darwin is credited with the theory of evolution because he published before Alfred Wallace, even as both were developing their ideas at the same time. As the need for publishers has actually declined in an increasingly digital knowledge commons (Hess and Ostrom, 2011), prices have not. When Aaron Swartz released hundreds of thousands of academic journal articles from the JStor aggregator platform (which universities individually pay huge amounts to each year), he was arrested and threatened with 35 years in prison, causing him to commit suicide in 2013, aged 26. Swartz had also been an avid contributor to Wikipedia almost from its inception. Tapscott and Williams (2008) make the interesting observation regarding knowledge production that the Wikipedia approach to peer-produced knowledge may produce more errors per entry on inception, but that over time corrections are faster and more detailed, so that Wikipedia in the end creates more accurate entries than more traditional rivals such as the Encyclopedia Britannica. The mapping of the human genome is just one example of where researchers depended upon the free circulation of the results of others around the world, and the ability to use such findings in developing their own work further (Sulston and Ferry, 2009). The subsequent use, however, of the work of the Human Genome Project by commercial actors who sought to patent particular elements of the overall genetic mapping process was both parasitic on freely shared work and damaging to its future development.

The kind of gift economy (of largely free giving of written works by academics in exchange for citations, which can be traded for career advancement) found in academia is celebrated, but also creates powerful hierarchies and cliques (Fuller, 1997). While Bronisław Malinowski's (1922) study of gift exchange in the Western Pacific Trobriand Islands showed a relatively flat circulation back and forth between equals in the Kula Circle, Marcel Mauss' (1925) account of Potlatch shows a more hierarchical model of redistribution by which the most powerful individual asserts their authority through the act of disposing of their surplus. Nick Davies (2008) argues that professional journalism today is caught between increased corporate concentration, on the one hand, and the circulation of opinion via new media, on the other. He refers to churnalism as the circulation of opinion within the media rather than the more expensive investigative journalism that he considers the true vocation of the professional reporter. The financing of news reporting is squeezed as revenue streams divert to online advertisers, and states are increasingly willing to deregulate media and hence reduce funding to public services. News production is increasingly integrated within global cross-platform infotainment corporations. Citizen journalists, and the information they offer up, may seem like a gift to the beleaguered professional journalist, as do the sound bites and press releases that well-funded lobbies and publicists flood news desks with every day. Both sets of gifts are offered freely, but Davies argues such gifts threaten the integrity of professional journalism and its vocation to deliver up something more than a catalogue of free gifts delivered up in the first instance by those with an axe to grind.

3. Tribes, bubbles and filters: then and now

Finding your tribe online and embedding yourself within that community is liberating for many, isolated in everyday settings. However, these silos may themselves divide society and fuel conflict. Also, it should be noted, yesterday's daily newspapers were far narrower, and readers more loyal to their blinkered sources, than are today's 'disloyal fake news audience' (Nelson and Teneja, 2018).

The term 'fake news' was, for a long time, used to refer to satirical, comedic parodies of news and news reporting, such as appeared on radio and television (Holbert, 2005), as well as in print. Edson C. Tandoc, Zheng Wei Lim and Richard Ling (2017) offer a six-part typology: news satire, news parody, fabrication, manipulation, advertising and propaganda. These are all based on the primary features of lack of 'facticity' (non-truth) and deception (not all untruths pretend to be true). The extent to which 'self-branding' by micro-celebrities evades the standards of truth expected in explicit advertising is a thorny tangent to this discussion (Khamis, 2017). The term 'fake news' has only recently, in particular since the 2016 presidential election in the United States, been applied to refer to news content circulated by means of new-media platforms and channels, which are deemed to lack the editorial authority of traditional news sources and content. It is, therefore, worthwhile looking at research into the power and significance of 'fake news', in both its traditional and newer meaning, in addressing the question of whether such content is having an impact on public beliefs about political issues and about politics, politicians and public life in general. Is 'fake news' corrupting public life? Well, the evidence seems to suggest that it is not.

Meital Balmas (2014) applies the more traditional conception of fake news to examine its relationship to trust in traditional news and trust in politicians and politics (in terms of belief in personal political efficacy, of alienation with the political process and of cynicism about politicians). The study is ingenious for its simplicity, examining viewing behaviour in relation to 'hard news' (defined in terms of traditional media newspaper coverage) and 'fake news', in the form of time spent watching a particular satirical news parody programme, Israeli television's *A Wonderful Country*. Balmas found that there was a very significant relationship between the amount of time spent watching 'fake news' (the satirical programme) and 'perception of fake news as realistic' (that is, the view of politics as pointless and politicians as cynical); however, an even stronger relationship was found between levels of time spent watching 'hard news' and not believing that fake news was realistic. Overall, the interaction effect of watching both hard and fake news produced a strong and statistically significant perception that fake news was not realistic. Those who were most exposed to fake news and, at the same time, least exposed to hard news had the most cynical, alienated and powerless attitudes, and saw fake news as most realistic; but this was a very small minority of the sample. Most viewers watch both hard news and satire, and for them the former largely cancelled out the influence of the latter. Tandoc et al (2018) found, in a sample of Singaporeans, that today's greater array of unreliable content online challenges internet users' capacity to judge content, but, when challenged, most users seek to compare claims made with alternative sources. Closure within tightly filtered bubbles is not common. Grace Craigie-Williams (2018) highlights a number of instances of 'fake news', including the case of a Muslim woman on her mobile phone, who was photographed after the Westminster Bridge terror attack in London (England) in 2017. Someone uploaded the image and claimed it showed the woman walking past and unconcerned, under the headline: 'Muslim woman pays no mind to the terror attack, casually walks by a dying man while checking phone … #BanIslam'. Craigie-Williams notes that the post was traced to a Russian hacker account, and the woman, it was later confirmed, was in fact ringing the emergency services. While the initial post was clearly 'fake news', and very likely deliberate disinformation rather than misinformation (being posted to look like it was from Texas rather than Russia), it was also the case that the hoax was soon revealed and circulation of this revelation is now greater than the original hoax.

The question then arises whether, in relation to what is today called 'fake news', the circulation of material online that is said to lack the authority of traditional journalistic and editorial controls follows a similar pattern to what Balmas found in relation to satirical content. The answer would appear to be yes. Jacob Nelson and Harsh Taneja (2018) used ComScore data for 30 fake-news sources and 24 sources of traditional news, mapping levels of online access in the United States in the period between January 2016 and January 2017 (the period of the 2016 presidential election won by Donald Trump). The results confirmed the 'audience availability' thesis – which is that low- to medium-level media consumers (in terms of time spent engaged with media content) tend to consume the most traditional content, while those who spend the greatest amount of time engaged with media content (old and new) also watch mainstream content most of the time. In other words, those watching non-mainstream content tend to be those that spend the most time engaged with media, and even they tend to spend most of their time engaged with the most popular and well-known programmes and channels. This is the traditional 'audience availability' thesis, but Nelson and Taneja found it fully predicts online media consumption as well. ComScore data showed that 'real-news' sites (such as the *New York Times* online) were visited on average 40 times more often than 'fake-news' sites: 94 per cent of those who 'visited' 'fake-news' sites also visited 'real-news' sites, and the time spent on real-news sites was on average twice as long as on fake-news sites (nine minutes as compared to four and a half). The research also found that, of those who accessed fake-news sites, the amount of time per month they spent on Facebook was three times that of those who did not access fake-news sites. The difference in time spent on Google was almost as great, although time spent on Facebook was around six times greater than that spent on Google. Facebook was the feeder in offering fake-news links in most cases. In essence, the vast majority of social media users orient, for the vast majority of their news-related content, to mainstream news sources, while the majority of those who do access fake-news sites also (and to a greater extent) access mainstream content. Only a small minority of people gain their news mainly from non-traditional news channels. For those that do, most is directed to them by Facebook. Teenagers are a group most likely to get their news from Facebook, and are the group least engaged in politics (Marchi, 2012). Whether Facebook can be blamed for teenagers' relative disengagement with politics is questionable, as most teenagers cannot even legally vote in elections. Politicians in democracies orient towards voters, so attention to the electoral system might be more useful than blaming social media for relative disengagement. Overall, the problem of 'fake news' is limited, and is, for the most part, something that is almost entirely dependent on one platform, Facebook, and so can be almost entirely dealt with at the level of that one platform's policy regarding what it does and does not allow/link users to do. Nelson and Taneja, however, include Fox News among its list of 'real-news' sites. What counts as 'the mainstream' in the United States may be seen as a source of 'fake news' in other countries. This should shift our attention to the nature of mainstream media in relation to political bias and the politicization of news. At least Facebook is, albeit under pressure from public and politicians, attempting to address its responsibilities for the circulation of what amounts to 'fake news'. Whether supposedly mainstream channels, like Fox News, take such responsibilities seriously is another matter. Eli Pariser (2014) highlights the need to be aware of how Facebook (and Google, Yahoo and so on) personalizes our feeds based on what we have previously looked at, creating what he terms 'filter bubbles'. We should be concerned, and we are, more and more. That we are not giving the same amount of concern to mainstream television (Fox News, for example) and newspapers (often even worse that television) when it comes to such bias, however, is problematic. Pariser is right to say that new media has not abolished gatekeepers, only replaced editors with algorithms. We should be symmetrically concerned.

Symmetry also manifests in the way new-media 'citizen journalism' may, in fact, mimic features of traditional journalism, in terms of whose voices are heard, and who brings them to us. In a study of citizen journalism's coverage of the 2010 Haitian cholera epidemic, those with access to the resources to upload citizen-journalistic content online often reproduced the same accounts of Americans as heroes and of Haitians as backward, as was the dominant framing of events in mainstream media (Krajewski and Ekdale, 2017). It should also be recalled that most self-declared 'citizen journalist' websites have an editorial agenda and editorial staff curating and managing the presentation of content, even if they do not always recognize that they are engaged in editing, and therefore do not always take responsibility of the biases they introduce or allow into the content they present (Lindner, 2017).

Diversifying those able to access an audience may or may not incite harm, even if a balanced and interconnected ecosystem of old and new media is better than either old monopolies or new echo chambers each concealing hidden actors and agendas.

Jinling Hua and Rajib Shaw (2020) suggest in their account of the first three months of the COVID-19 epidemic in China that the management of information was key to containing the spread of the disease. This involved the government exerting their control, both over information about citizen movements, so as to be able to contain population movement, and over what was considered 'fake news' about the epidemic, so as to prevent panic. China's relatively rapid and successful early containment of COVID-19 can, at least in part, be attributed to the ability of the government to contain information: that is, messages contrary to their policy of containing the virus, as well as any criticism of their response regarding the origins and spread of the disease in its early stages. Hua and Shaw do not name Dr Li Wenliang, only referring to the death of a doctor who had posted critical comments online. That he was arrested, and his criticisms deleted, evidences the power of the Chinese government in controlling what they saw as a threatening 'infodemic', but while state control of media prevented panic early in the pandemic this containment was not helpful in the long run. COVID-19 deaths in Western countries were much higher early on and overall, how much can this be attributed to weaker state control over information? China's later failure to contain the virus by authoritarian means suggests control over information and bodies creates significant limitations, even as did forms of individualism in the West that refused to enact or even believe rules and expertise handed down by authorities.

4. More speech is better than less

New media does show that more information is better than censorship, even if not all new-media content is honest or factually correct. A dialogue between professional journalism and citizen journalism needs to take place. Tony Harcup and Deirdre O'Neill (2017) set out to update their earlier typology of 'news values' for traditional journalism in an age of new media, where audiences get to engage and reply to what professional journalists seek to define as news. In earlier work, these authors identified ten core elements within a prospective story that likely made it 'newsworthy' to mainstream/ professional journalists: involvement of the power-elite (economic/political); celebrity; entertainment; surprise; bad news; good news; magnitude of consequence; relevance to existing news stories; follow-ups to existing news coverage; and association with existing editorial agendas (most often driven by 'commercial news values' – advertisers, proprietorial interests or other vested interests able to influence editorial agendas). The dominant tension within this typology is between a focus on the status quo (power elites, celebrities, entertainment and editorial agendas, along with things relevant and following up from these), and 'bad news' that involves some breaching of the elite agenda (whether in terms of conflict or disasters). 'Bad news' was the largest single 'type' of news, even if the sum of elite-oriented news (political, economic, celebrity and entertainment elites) amounted to far more coverage. Harcup and O'Neill note that news stories in Facebook feeds (and the UK's most popular online newspaper site – the conservative *Mail Online*) tend to be driven by conflict. However, these stories are more often in relation to 'entertainment' than to traditional bad news, even if 'conflict' is also a key component of what makes mainstream journalistic 'bad news' (when elite consensus is challenged/disrupted). At least in terms of audience engagement with mainstream news via online feeds, content is even more likely to focus on entertainment (rather than challenges to power elites) than is the case with mainstream news (even if the latter's attention to such disruptions tends to be largely supporting of the status quo). The shift in news, from political and economic power to 'human interest' stories, carries on apace in terms of Facebook feeds, but this is a continuation of a trend already present in the history of mainstream news in any case (Habermas, 1962). Nonetheless, the influence of Facebook is significant in extending this focus, not just in what its users read, but in terms of what mainstream journalists write. Harcup and O'Neill (2017, p 1475) cite Emily Bell, who claims: 'The key question for news organizations, tied to the goal of big traffic, is now "what works best on Facebook".' Conflict fuels clicks. Even if most of this conflict-oriented 'click bait' is relatively superficial entertainment gossip, the same algorithms will deliver polarizing political stories to those with a predisposition to respond to it (even if that is a relatively small number). Kirsten Johnson (2018) looked at over 500 articles from US online citizen-journalistic websites. Most citizen journalism replicates mainstream content and values. Carpenter, Nah and Chung (2015) found a disproportionate number of graduates and around 60 per cent of citizen journalists were male, which may bias reporting and topic selection.

Timothy Garton Ash (2016) refers to a development beyond 'the daily me' (algorithms that feed existing, filtered preferences) to 'the daily kiosk', where new-media users upload into their bubbles, as well as just feed within them. Nevertheless, he argues, the capacity to use new media to widen the discussion about the future of humanity remains and must be developed. If 1989 gave us the collapse of the Berlin Wall and the invention of the World Wide Web, it also saw the Fatwa issued against Salman Rushdie (for his book *The Satanic Verses*) by Iran's then supreme leader Ayatollah Khomeini, and the crushing of China's pro-democracy Tiananmen Square protests (in which thousands died). Garton Ash suggests the need to create something akin to Habermas' ideal-speech situation ('inclusive and noncoercive rational discourse among free and equal participants') is all the more essential in a globally interconnected world, and all the more possible because of such connectedness. That is not to say this will be easy. In particular, the relationship between state and business interests, professional journalistic values and non-professional citizen journalism has to be organized so the value of professional standards and that of citizen voices can be combined in order to limit the power of states and business interests. This is even as the boundary between free speech and hate speech means states still have some role to play in limiting expression. Still, the right to disagree, even to the point of giving offence, can, Garton Ash argues, be defended as a universal human right. This is the case against regimes, institutions and individuals who claim the right to silence those who question them, and against the imperialist legacies of European powers that themselves claimed always to be acting in the name of universal values (when they rarely were). Given Harcup and O'Neill's findings regarding Facebook feeds, and Garton Ash's concerns about current attempts to curb free speech around the world, do citizen journalists successfully make a positive contribution to creating a better-informed world? Seong-Jae Min (2016) suggests this conversation between professional and citizen journalism can and is taking root. Jing Zeng, Jean Burgess and Axel Burns (2019) examined Weibo users in China after the 2015 Tianjin blasts (in the north of the country). China's severe criminal penalties for spreading rumours online (up to seven years in prison), and these penalties applying to service providers like the Weibo platform as well as its users, are designed to limit challenges to official policy and reporting of events. However, a form of citizen journalism on Weibo disseminated all the early filmed coverage of the events, and released information about the deadly chemicals involved, the risks to firefighters of chemicals that would react to water, and reports of the high death toll among the firefighters sent (ill-informed) to deal with the fires. Official and social media reporting was heavily censored, but information and questions did circulate, and the way citizen journalists checked facts and reported them far outstripped the official reporters in their adherence to what are commonly assumed to be the values of 'professional' journalism.

Joy Leopold and Myrtle Bell (2017) examined mainstream newspaper coverage of the Black Lives Matter (BLM) movement – in particular, content covering events after the 2014 killing of Michael Brown by a police officer in Ferguson, Missouri. Leopold and Bell found that mainstream media coverage reproduced what they call the 'protest paradigm', where critics of the status quo (in this case, the police and the justice system that decided not to press charges against the officer that shot Brown) are presented as problematic and not to be believed. Chloé Cooper Jones (2019) records how filming the death by choking of Eric Garner by a police officer in New York failed to secure a conviction. Ramsey Orta, who filmed the killing on his phone, was himself arrested and imprisoned. However, the film footage did get out, and while the officer was never charged with the killing, the family did secure an out-of-court settlement with the New York Police Department, and the officer was (five years after the killing) fired from the force. The struggle to make visible the invisibility of violence against black men by the police in the United States (Powell, 2016), even with the aid of filmed footage, has met constant and significant resistance from the police and mainstream media. Where Ramsey Orta's voice and even his filmed footage was discounted, the multiple recordings and immediate uploading of footage online (and especially via the #BlackLivesMatter website) of the killing of George Floyd in 2020 was not able to be made invisible. The conviction of the white officer, Derek Chauvin, who killed Floyd, and his being sentenced to 22 years in prison for murder, would not have happened if not for the footage circulating online and the BLM movement and website (as an activist form of citizen journalism) to promote that circulation. That mainstream media were then willing to take up the story is also a significant dimension of a changing overall media ecology. Outside the United States, citizen journalists using social media have been ahead of official reporters in challenging the authorities in repressive regimes, including Egypt (Mrah, 2019), Zimbabwe (Moyo, 2011), Myanmar (Pidduck, 2010), Russia (Simons, 2014) and China (discussed in this section).

5. Why the media loves to blame the media

Blaming the media for social polarization is the knee-jerk response of politicians, and blaming the new media is something that politicians and traditional media leaders can find common cause over. While fake news can, and often is, fuel for an emotional economy of blame that reduces scope for dialogue and increases hostilities between groups (or perceived groups), state and commercial media are no less guilty of such ideological stoking. Vladimir Putin did not win the 2016 US presidential election for Trump by spreading fake news, even if Rupert Murdoch cannot take all the credit either (despite the latter having the stronger claim). Fake-news websites do not exert significant influence beyond a limited and predisposed, partisan audience.

Traditional print and broadcast media is often guilty of being ideological, in the interests of state and business elites. What Herman and Chomsky (1995) refer to as the US-propaganda model of mass media is replicated (although distinct in the reflection of context-specific state/business elite formations) in China (Garton Ash, 2016), Russia (Pomerantsev, 2016; as well as Khadarova and Pantti, 2016), the United Kingdom (Glasgow Media Group, 1976) and across Europe (Habermas, 1962). Is fake news online better or worse, in extent and/or impact? New media does increase the capacity to exercise free speech as never before, but does free speech also mean freedom from any association with truth (Dorf and Tarrow, 2017)? Vosoughi, Roy and Aral (2018) found that online rumours found to be fake circulated far more widely and faster than rumours that were true, because fake news tended to be making bolder (exaggerated) claims that triggered more emotion and a greater tendency to forward that 'information' to other (assumed to be like-minded) people (humans being more prone than robots to forward exaggerations).

Vargo et al (2018) examined the 'news-agenda-setting' (NAS) power of online fake-news sites in the three years running up to the 2016 presidential election in the United States. They conclude (2018, p 2038): 'it appeared that fake news did not set the overall online media agenda for those years examined; instead, its NAS power turned out to be shrinking over time'. Overall, fake-news site content's influence was weak in predicting topics and linkages between topics in mainstream media. It was partisan media channels (both online, such as Breitbart, and traditional, such as Fox News and conservative talk radio) whose topic coverage and linkage set the agenda for subsequent fake-news stories trending online. Liberal-oriented, partisan media outlets were more likely to follow fake-news stories trending online (which may often mean engaging claims they dispute) than were conservative channels, who were more powerful in setting such trends. In essence, it is the existence of a highly partisan media landscape in the United States, not the rise of fake news online, that is the driver of an increasingly polarized news agenda in that country. Michael Butter (2020, pp 130–136) highlights the contrast between this highly partisan media landscape in the United States relative to the situation in Germany (and, to a degree, elsewhere in Europe). Contrast the few thousand conspiracists linked to Germany's Reichsbürger movement with the tens of millions of people linked to QAnon in the United States. Both used digital networks to access and incite, even as their failure to evade and conceal has enabled authorities to monitor both. The latter was huge. The former was much smaller. Where around a thousand people were arrested for their part in the actual storming of the US Capitol building in January 2021, only a couple of dozen arrests were made regarding an alleged plot (never enacted) to do something similar in Berlin. In Europe, containment within filter bubbles is very much more limited than in the United States, and (as discussed in Section 3) such containment is also relatively weak in the United States. Even in the United States, most news gained online (even when filtered through Facebook) comes originally from mainstream news channels. Vargo et al (2018, p 2028) are therefore right to argue that, even as, 'content from fake news websites is increasing, these sites do not exert excessive power'.

Nevertheless, small but highly antagonistic extremes may generate significant political hostility and potentially violent confrontation. Over time, the decline in advertising revenues (and state funding) for traditional news production, and the need to chase clicks via Facebook and Google, may produce a shift to what Bakir and McStay (2018) call the economy of emotions. For these authors, Facebook's 2010 newsfeed algorithm Edge Rank, which determines what users receive in their newsfeed based on the combination of prior search history and current advertiser funding priorities, has replaced news editors with algo-journalism (where algorithms select content and increasingly come to write it too). In a polarized environment, and with significant partisan media presence, the drive to generate clicks through highly emotive headlines and storylines, and the ability to identify and target users through knowledge of past 'likes', make Facebook's business model prone to promoting content that puts conflict ahead of accuracy. Bakir and McStay document a range of post-2016 strategies by which Facebook has sought to limit its role in the circulation of fake news: content monitoring; third-party verification and fact checking; warning labels; developing a more editorially based policy in terms of prioritizing/deprioritizing content; blocking; and removing spoof domain names and so on. The authors conclude that a more powerful approach would be to target advertisers and make them more responsible for the content that their advertising is linked to. Just as Facebook itself is increasingly being held liable for what users post (as editorial media is for what it publishes/broadcasts), if advertisers were held liable for what their ads are linked to, revenue generation from fake-news-content 'clicks' would very quickly collapse. To a significant extent, this is now the case. Ironically, the capacity to reform one platform, Facebook, is far easier than reforming the raft of traditional media platforms whose cumulative impact in promoting propaganda might be far greater. Hal Berghel (2017) notes that Facebook's five most-clicked-on 'fake-news' stories during the 2016 US election were pro-Trump/anti-Clinton stories, illustrating the scale of bias in user engagement. Still, that Facebook measures these things means it was able subsequently to enact policies to reduce their repetition in the 2020 election, which Donald Trump lost (not that he got more votes than Hillary Clinton in 2016 either). At least Facebook creates the 'paper trail' that a more robust fact-checking and blocking policy can then follow up on (and largely has now). Axel Gerfert (2018) is correct in suggesting that deliberate disinformation (designed to mislead) is both more problematic than simply the spread of misinformed opinion and more readily identifiable (due to the way it has been designed). In essence, such content can be relatively easily blocked on Facebook if the choice is made to do so. Of course, to the extent that not all claims can be simply and objectively defined as true or fake, 'opinion' should not be considered a problem as such. Whether we trust Facebook or any particular government to make such decisions is, then, the issue, rather than it being the nature of technical networks as such. It is not simply the case that populations have become more willing to believe what they read today, or that they are increasingly sceptical about everything they read. There is increasingly partisan belief among some (Vargo et al, 2018) and greater doubt among others (Mould, 2018). Most people retain a balance between the two, just as most people retain a balance between new and traditional media news consumption.

It was the *New York Times*' (eventual) willingness, in 2017, to publish allegations of rape against Harvey Weinstein (Kantor and Twohey, 2019) that led to other victims coming forward, and for the (until then) relatively unknown #MeToo platform to become the globally significant forum for reporting sexual assault that it became. The existence of online spaces like #MeToo (as was the case with BLM and WikiLeaks) also generate and circulate evidence that can then reach a wider audience through mainstream media. The overall media ecology created by the combination of traditional and new media has, predominantly, created a more informed society.

In conclusion

As hacking scandals involving mainstream media companies illustrated (see Section 1), it is not simply new-media hackers that can access citizens illegally. The rise of the global network society has seen increased mainstream media concentration in terms of vertical, horizontal and cross-national integration, even as digital networks have afforded more distributed forms of online communication. Access to audiences by increasingly concentrated media enterprises and by distributed media forms has increased, just as access by audiences extends in both directions. Behind a wide array of imprints and channels, mainstream media concentration is increasingly concealed. Most of what audiences consume online is produced by a narrowing range of mainstream sources (of entertainment, news and infotainment). That disinformation-oriented, fake-news generation can conceal itself within (typically) Facebook newsfeeds is increasingly limited by Facebook's own policies, although such content continues to exist and circulate. While Facebook's Edge Rank newsfeed algorithm has, since 2010, been the vehicle that such fake-news generators have sought to exploit in order to generate clicks, it is also a singularly powerful vehicle for blocking such practices when sufficient pressure is brought to bear on the company. Evasion of such blocking measures remains an issue precisely because the balance between overly draconian blockage of rumours might be considered worse than more limited actions that allow elements of disputed material to circulate that some, but not all, would consider false. The key issue is the extent to which fake news incites criminal harm. As this chapter has documented, fake news online has limited impact, and is for the most part discounted by most viewers, even if for some capture in echo chambers can escalate insular relations into hostile actions. Filter bubbles and echo chambers are largely offset by the majority whose news consumption contains mostly mainstream content (even when this is mediated by online platforms). The work of Maria Ressa (2022), spanning the boundary between mainstream journalism and the creation of her own online platform – Rappler in 2012 – won her the Nobel Peace Prize in 2021. In a polarized society, with highly partisan mainstream media, fake news online can incite further hostility; but it is society, not new media, that is the primary driver, and policies to moderate online content are readily available precisely because of the overwhelming significance of one player (Facebook) in circulating news (and fake news) online.

PART IV

Appropriation

8

Fraud, Extortion and Identity Theft

Key questions

1. Does the rise of online fraud and extortion indicate increased risk or simply the migration of already existing forms of crime to online domains? Or is this rise just the consequence of economic growth, particularly in intangible goods?
2. What is identity and what does the concept of identity theft tell us about the relationship between identity and identification in a digital age?
3. Why is digital encryption so secure in some domains but so weak in others?
4. What is the relationship between social engineering (persuading people to give up their personal details) and technical skill in online fraud?
5. Is the hidden extent of fraud online greater or less than in relation to older forms of fraud?

Links to affordances

The issue of incitement is significant in relation to fraud and identity theft (insofar as people are persuaded to act on the basis of deception), while the question of access is paramount. The increased separation of person from persona, identity from identification, in and by means of digital networks, makes alienable identifiers both increasingly essential and risky; but personal safeguarding and institutional security (and compensation) methods make most people safer online than off (although this may itself incite risky/ unthinking behaviours). What encryption (and the digital keys that can be used to secure/unlock encrypted data) can conceal, surveillance or poor safeguarding can *reveal*, but this creates symmetry between fraudsters and authorities seeking to follow the digital mouse droppings that online fraud itself creates. While evasion by means of transnational access is made increasingly possible, the social-engineering side of networked fraud limits such action at a distance.

Synopsis

The possibility of micro-fraud, where small amounts are taken from large numbers of victims, increases in scope online, not least as more people use ever greater numbers of online platforms to undertake financial transactions that they would previously have either undertaken in a face-to-face context or not at all. This has led to a huge escalation in the incidence of online fraud. In addition, more traditional forms of large-scale fraud are also made easier through digital channels to digitally held accounts. As such, fraud online has increased, but the question remains whether this manifests an increase in overall levels of fraud or a migration from one modus operandi to another. This chapter highlights a complex picture. Overall levels of fraud are on the increase because overall levels of financial transaction have increased over time, and overall levels of online fraud reflect this, as well as representing a migration away from financial exchange and fraud by non-digital methods. (All of these points apply in relation to online extortion [ransomware] as well.)

Scope to steal a person's identity, in the sense of appropriating their forms of identification, increases online precisely because online forms of identification must, by the very nature of transactions being 'at a distance', be more 'alienable' (separable from the embodied person) than traditional physical acts/sources of identification. The separation of identify and identification is intensified in online domains, even if the distinction between personality and persona (essence and appearance of self) has far longer roots. Still, while 'identity theft' is significantly escalated online, capacity to take over someone else's identity by digital means has its limits. Encryption is never absolute, but in transactions between institutions and individuals where both parties have an interest in maintaining confidentiality, digital keys are, in many ways, more secure than physical ones. Where sharing keys with non-trusted others takes place, digital encryption falls apart (as will be seen in Chapter 9 on 'sharing').

Chapter sections

1. Online fraud is increasing rapidly. In part, this is migration online of older forms of real-world fraud, simply because the digital is where transactions increasingly occur (Wall, 2010). Certain behaviours significantly increase risk (Pratt et al, 2010). Other, precautionary, actions reduce vulnerability (Bossler and Holt, 2009a). However, actions of banks, shops and platforms to protect customers, such as compensating victims, may increase risky behaviours. Online banking and shopping are relatively safe. One significant danger is to believe that such transactions are relatively safe (even though they are).

2. The term identity theft is relatively new, describing something that takes on an unprecedented significance in an age of digital network mediated actions. The separation of identity and identification is a precondition of online financial transactions. This creates an intrinsic vulnerability – sharing with strangers what secures you from strangers (Wall, 2013).

3. Encryption is designed to secure identification methods, but sharing data with strangers makes such security a problem (Cross, 2015). Nonetheless, forms of digital encryption are relatively secure (by comparison to physical keys, which any professional locksmith can bypass in seconds), as long as you only share key codes with those that have the same interests as you – your bank, for example. Sharing encryption keys with non-trusted others creates vulnerability. As such, online banking remains much more secure than, for example, the encryption placed on digital goods such as music, films and software – all of which are broken as soon as they are circulated (Bossler and Holt, 2009b).

4. The classic confidence trickster is today called a 'social engineer' (Reyns, 2013). Far from the separation of the social engineer and the technical geek, the most successful digital fraudsters have strong social skills and/ or other social means of accessing personal information (Rege, 2009).

5. The problem of knowing how much or how little we know about the true extent of fraud and identity theft online will not go away, but at least it is possible to map the causes of our ignorance (Cross and Bradshaw, 2015; Levi et al, 2017).

1. Follow the money?

While the rate of growth in online fraud has been high, a large part of this is explained by migration to online fraud from older forms of fraud, simply because that is where transactions increasingly occur. Certain kinds of online behaviour radically increase risk, as other, precautionary, kinds reduce vulnerability. However, measures to reassure potential online customers of banks, shops and platforms, by promising to compensate victims, may increase recklessness. While online banking and shopping are relatively safe, one of the greatest dangers is the belief that such transactions are relatively safe (they are).

In 2012, the US Internet Crime Complaint Centre (ICCC) recorded $525 million in online fraud losses, but this was estimated to be only a small fraction of the total lost. Its estimation of total losses was $24.7 billion (Clough, 2015). The ICCC recorded 17,000 complaints in 2000, 231,000 in 2005 and 314,000 in 2010, with subsequent rises but of a less steep level. The vast majority of these complaints were regarding fraud, with much smaller numbers relating to hate speech, cyber stalking, abuse, grooming and child pornography. According to UK Office of National Statistics data

(ONS, 2016), between 2000 and 2010, the financial value of overall reported fraud fluctuated up and down, but ended only a fraction above where it had begun, with great spikes and drops in between. Adjusted for inflation, the overall increase may have been nothing at all. The scale of fraud online on the other hand, while showing some fluctuation as well, ended up being around 50 times higher at the end of the decade than it was at the beginning. As a proportion of overall recorded fraud value, online fraud rose from being 1 per cent of the total to being 37 per cent. According to stopfraud.org, over the course of the following decade (2010–20) online fraud in the UK doubled again. If overall levels of fraud remain static, while online fraud increases, that does rather suggest that the internet is not increasing fraud as such, but only constituting a greater part of what fraud there is. According to the UK Office for National Statistics again, Crime Survey for England and Wales data suggests that the proportion of people who fall victim to online card fraud each year rose in the decade to 2010, from 3 to 6 per cent per year. However, since 2010, it has fallen back to, and remains static at, just under 5 per cent (ONS, 2016).

David Wall (2010) argues that, of all crimes, fraud appears to be the one that has taken to the internet most successfully. For Wall, the internet is a force multiplier in the extent to which multiple victims can be targeted remotely (access), and with far less risk to the perpetrator relative to face-to-face robbery or fraud (evasion). The capacity to skim small amounts from large numbers rather than large amounts from singular targets also radically increases the capacity for such online frauds to go unnoticed (concealment). Credit card fraud in pre-digital payment times saw counterfeit cards passed off as real; but the rise of network confirmation of payment (by 'chip and pin') made such counterfeit cards easily identifiable, leading to the rise instead of cloned cards (where real card details are skimmed and copied). With the escalation of online purchases, not just the confirmation of payment in shops by electronic means (chip and pin), the scope to engage in 'card-not-present' fraud has escalated. Gleaning a person's personal information by hacking into databases or online communications in order to gain credit is another dimension of online fraud. Wall distinguishes between virtual bank robberies, where banks, shops and customers have money (or goods via credit) taken from them under false pretences by means of fraudulent use of victims' accounts, or through the creation of fraudulent accounts gained with the use of victims' personal information. Beyond such account/card deceptions, online stings use arbitrage, false advertising online, premium call charges and short-firm fraud techniques, while online scams operate virtual pyramids, investment hoaxes, entrapment, 'scareware', false auctions, advanced fee tricks and so on, promoted and enacted online, to defraud remote victims. The scope to access, conceal and evade online is huge, but the scope to incite, to persuade victims to give out personal information or to pay for something in error, has its limits. Still, what distinguishes fraud from robbery is that the victim is persuaded (incited) to hand over money – whether that be a bank persuaded to hand over a customer's money because they believe the request for payment has come from that customer (and not someone impersonating them), a shop likewise prepared to give credit to someone they believe to be someone they are not, or an individual or organization persuaded to pay for something they will never receive. Fraud requires that the victim is incited; but despite the force-multiplying capacity of digital networks in terms of access, evasion and concealment, most people, most of the time, are not victims.

Adam Bossler and Thomas Holt (2009a) sought to answer the question of what it is that best predicts why some people are victims of online crime (including fraud) while most are not. More specifically, they sought to test the 'low self-control' hypothesis, which suggests that victims are more likely to engage in risky behaviour that then explains why they are more likely to be victims. After controlling for other variables, these authors found that this trait had no predictive strength in relation to the likelihood of being a victim of online credit card (virtual bank robbery) fraud. Even if individuals with lower levels of self-control were more likely to reveal password information, and were more likely to experience their computers being hacked and online harassment, they were still relatively robust when it came to keeping their financial information secure. However, it should be noted that other personal information can be abused to build up a set of details to be used to fraudulently gain credit in someone's name. Also, if low self-control does not predict credit card fraud, this does point to the fact that everyone is a potential victim if they use such cards, especially if they use them to make online, card-not-present, payments, where it is harder to verify the person with whom you are making a transaction. Travis Pratt, Kristy Holtfreter and Michael Reisig (2010) looked at these behavioural predictors, rather than personality-based predictors (which, as noted, were not very useful). It was the amount of time spent online, and whether or not you had made a purchase online, that predicted likelihood of online victimization, not demographic/personality features. Although the young and the educated were more likely to be victims, this relationship disappeared when researchers controlled for the fact that these two groups are simply the highest users of the internet (and of online purchasing, in particular). Being available to be victimized (in terms of time and making purchases) predicts victimization, yet most users are not victimized most of the time.

2. Identity theft

Identity theft is a relatively new term (coined in 1964 and popularized only decades after that – see Poster, 2006) to describe something that, while not entirely new, takes on an unprecedented significance in an age of action at a distance: the need to separate oneself from one's means of identification. The separation of identity and identification, the alienability of such properties that are 'taken' to be inalienably one's own (and hence prove a person to be themselves), is a precondition of online financial transactions, but one which creates an intrinsic vulnerability – sharing with strangers what secures you from strangers.

David Wall (2013) locates online identity theft within a threefold typology of identity crimes: first, theft of personal data (or, rather, legally speaking, the unauthorized access and use of such data) that can act to identify a person; second, the creation of false identities for the purposes of fraud (that may involve creation of fabricated identities or the use of real others' identifiers); and, third, the use of such false identities to commit crimes by means of fraudulent identity claims. These crimes may involve false identities being used to defraud others, or to undertake forms of extortion or intimidation while concealing the identity of the perpetrator/s. Extortion using fraudulent identifiers may be designed either to conceal the identity of the perpetrator while still directly threatening the victim (direct extortion), or to create the impression that the perpetrator represents an agent seeking to assist the victim in retrieving lost property or access to their computer. Wall outlines a second typology of identity theft as a precursor to fraud (and extortion by means of deception). Where 'real-world' means of gaining such identification data exist (by various forms of trashing/dumpster diving and social engineering: persuading people to give up information), digital networks have acted as a huge force multiplier in terms of access, concealment, evasion and (through persuasion) incitement. En-masse emails and texts inviting prospective victims to engage in one or another kind of false interaction/venture (including romance scams, investment schemes and publishing opportunities) abound, but such *phishing* schemes have very low rates of reply and tend to put most recipients on their guard. Still, such schemes can be continually 'reskinned' to play on whatever is the latest source of social concern (such as with COVID-19-related scams in recent years). Spear phishing uses personalized data in the emails that are sent out to create the impression that the message is from the prospective victim's bank, employer or similar. However, spear phishing requires a higher level of prior identity theft to initiate further intrusion in the attempt to access details that would then allow access to the victim's finances or other lucrative identifiers (such as might then enable the fraudulent application for credit or a passport in the victim's name). *Pharming, cache poisoning* and *spoofing* involve the creation of counterfeit websites, which victims are directed to (if their computer has been infected with a virus that makes this switch when the user seeks to access their legitimate bank or service provider). This is a more sophisticated mode of deception, as it bypasses the potential victim's active choice to follow an instruction or open a message. Just as chip-and-pin technology was developed to offer an online solution to the problem of counterfeit credit cards, so the development of dual-factor authentication (where secondary verification by means of a user's mobile phone is required in order to confirm their identity) reduces the capacity to defraud service users in the outlined ways. *Shishing* and *vishing*, where texts or voice calls are used to gain access to additional security codes (such as with card-not-present security numbers on the back of debit and credit cards), seek to outmanoeuvre emerging and additional security measures. Spyware and botnets have automated and escalated further the capacity for digital networks to harvest identity details (theft of such data) for the subsequent purpose of identity fraud (theft by means of such data). Wall notes that, while measures can be taken to limit the capacity of online identity theft for the purposes of fraud (and extortion), the affordances of digital networks challenge both the patch orientation of police routines (access at a distance) and the real-world primacy in policing of victim reporting. Identity theft online, and fraud online, create scope for high-volume/low-value theft. Many victims are defrauded of small amounts rather than single, large-scale thefts taking place. The scale of victimization on any one individual may be so low that either the victim does not notice or (taken on its own) it may be below a level that the police would routinely take seriously ('the law does not deal with trivialities'). Policing fraud and identity theft online, therefore, requires guardians pre-emptively looking for perpetrators rather than waiting for victims to report what they may not even be aware has taken place; but if perpetrators are acting at a distance, concealment and evasion makes such an approach difficult within local priorities and budgets.

Mark Poster (2006) found that media coverage in the United States of 'identity theft' was virtually non-existent until the mid-1990s, and still very rare between 1995 and 1999. Since that time, coverage and concern over identity theft has exploded, and this is, for the most part, the product of increased use of digital networks, where information is increasingly valuable not just as a means of access (identifiers as confirmation of identity) but also as a form of property in itself. Poster notes the distinction between property such as money and physical assets, and properties of personhood linked to notions of 'identity'. He documents the history of identity, from Locke's conception of identity as bound up with consciousness, through to early 20th-century conceptions of identity being bound up with the ego's management of desire in relation to the external world of necessities. Today, 'identity' is increasingly bound up with the management of representations of self – but, in a networked world, such representations and the management of them are increasingly complex, fragile, contradictory and fluid. The relationship between identity (as inner truth) and identification (as external representation), and, within the latter, between external representations as keys (personal ID) and external representations as affiliation to collective identity (group identity), becomes more and more complex as forms of identification become increasingly (of necessity) separated from the embodied person whom they are said to identify. Digital identifiers give access to more and more data (financial and medical) and detail (diaries and photographs). This content (which contains our identities) is held remotely, and the keys (our identifiers) have to be able to travel beyond our embodied person in order to access our content. Poster (2006, p 100) writes: 'The massive migration to the Internet of financial and business transactions, records and archives of institutions of all kinds, personal communications, diaries, and family memorabilia gives rise to a new realm of social and cultural life that may be hacked, intruded on, defaced and mutilated, appropriated or destroyed.' Just as the Industrial Revolution created a new mass of material things that could be stolen, so the recent revolution in information has created a new mass of immaterial things that can be stolen. These things are called 'identities'.

3. Diverse digital pathways to fraud

Fraud requires the incitement of a victim (whether that be an individual, institution or technical system) to act against their own best interests/ lawful design by means of deception. Digital networks afford greater access, concealment and evasion by which such incitement may be achieved. Encryption secures data but sharing data access weakens this. Digital encryption is relatively secure (relative to physical keys) if you only share key codes with trustworthy others. Sharing encryption keys with strangers creates vulnerability. Online banking is therefore more secure than, for example, digital copies of music, films and software – all of which circulate relatively freely.

Online fraud may involve end-user human deception, whereby the victim is persuaded to actively pay the perpetrator, having been tricked into believing the person defrauding them is offering them a genuine service or benefit. Alternatively, fraud may operate through the theft of personal information that is then used to defraud an institutional system into paying out or issuing credit in the name of the person whose details have been stolen. Online extortion, entrapment and scareware may also use the former type of deception as part of their mode of robbery, but not in all cases. Locking down someone's computer or gaining access to embarrassing personal information and demanding payment may or may not have involved prior fraud. Here, the two primary modes of fraud (deceiving the end-user through social engineering via concealment, and system deception through use of stolen personal data via remote access) are examined.

Cassandra Cross (2015) interviewed 85 Australian seniors about online fraud. Cross chose seniors because they are often the primary targets of online fraudsters, being assumed to be more susceptible to email-based, advanced payment/fee frauds, and more likely to have savings, pensions and other assets to be stolen. Most of those interviewed had never been confronted by online fraudsters, while, of the 28 who had, around half had been initially engaged with an attempt to defraud them but had disengaged after becoming suspicious; only a minority, then, had both initially responded and subsequently been defrauded. What Cross found was that among those who had not been defrauded, and even among those that had, there was a dominant narrative that blamed the victim for being both greedy and gullible in having been taken in by such fraudsters (Cross, 2013 and 2015). This tendency to blame the victim assumes that such frauds are easy to spot and that only someone very foolish and overwhelmed by the promise of some kind of unearned reward would be taken in. It is remarkable that even those who had been taken in came to blame themselves for their own supposed stupidity. Cross suggests that professional fraudsters are skilled in techniques designed to put a victim under pressure and to play on currently trending fears in their framing of the issues around which the victim is persuaded to make payment. Concealment and incitement remain central to what fraud has always been, even as digital access and evasion offer new avenues to such criminal deception. Most people are not subject to such victimization. Even though most of those who are approached will delete such invitations, a minority *are* tricked and they should not be condemned as greedy idiots, not least because this reduces the likelihood that they will come forward to report their victimization, and so increases the risk to others in the future (reporting rates being low already; United Nations, 2013). 'We always laugh at them' was one response that Cross received to a question about perceptions of victims; but then one of the respondents who had, in an earlier interview, laughed at those that would fall for such scams became a victim himself by the time of a later interview. There are plenty of things to look out for in relation to online fraud. Some degree of self-awareness around this issue is clearly going to significantly reduce the risk of being a victim. However, the person who thinks they are too clever to be a victim, and who therefore belittles others who have been taken in, is the person perhaps most likely to be a victim in the future. Increased awareness of advanced fee scams, romance scams and other 'reskinned' variations of these kinds of online fraud do make internet users less likely to be victims of them, but reskinning means ever new forms of disguise arise.

The person who thinks they would never give away their personal information, and in particular their bank details, should be aware that fraud by means of malware bypasses conscious awareness. Infecting a person's computer and thereby stealing their personal details to then be used to defraud them (or their bank/potential credit provider) does not require their knowing involvement in the sense of being tricked by means of social engineering (persuasion/incitement to cooperate). Again, it is easy to believe the person who allows their computer to be infected is somehow to blame for their own victimization, but this, again, is a mistake. Bossler and Holt (2009b) found that higher levels of protection afforded by using institutional computer networks, rather than relying on personal computer anti-virus software updates, did offer a greater level of protection from having one's personal data stolen through malware attacks by means of viruses accidentally downloaded onto one's machine. Ironically, it is users who rely on employer or educational institutional networks who tend to be less tech-savvy, while those who think they are more tech-savvy are more likely to self-manage and they are more likely to be victims of malware. Meanwhile, the belief that those who are victim to malware attacks are more likely to be engaged in illicit forms of internet use was only partially confirmed in Bossler and Holt's study. Users who actively engaged with certain piracy websites were found to have an increased likelihood of downloading malware. However, as the next chapter (Chapter 9) will point out, the leveraging effect of music piracy in the development of legal services that offer free access to music (such as advertising-funded streaming services) makes the use of illegal sharing sites largely redundant today anyway. Interestingly, Bossler and Holt found – based on respondents' reportage of their own online habits – that it was not accessing pornography online that increased risk of downloading malware, but, rather, interacting online with other people who *did*. This interaction increased the flow of emails from such illicit users and so increased the flow of potential malware conduits, that is, emails and web links containing viruses. However, it might be noted that respondents' associates accessing such sites did, in the first place, increase risk, and so rather implies that such sites are, per se, conduits for viruses to those accessing them – so perhaps those in the sample were feeling a little shy when answering the question about 'their friends' who watch pornography online. In conclusion, Bossler and Holt suggest that it is not strong computing skills and careful password management that reduces threat, but limiting one's access (first- and second-hand) to illicit networks of computer users, which most users do. As such, there is only so much you can do to avoid infection, and the most effective method is to have strong guardianship at the level of network and service providers. Again, blaming the victim is largely an erroneous assumption.

Bossler and Holt found that over a third of their (student) study participants had experienced some form of data breach in the preceding 12 months. Yet, as noted earlier in this chapter (see Section 1), the percentage of the population experiencing fraud annually is significantly smaller. According to UK Finance (2021), measures designed to stop online fraud have reduced its incidence in recent years, even as levels of online shopping have escalated. UK Finance estimate that action by banks saw two-thirds of account-based fraud stopped as it was being undertaken. However, while chip and pin was developed to prevent older forms of credit card fraud, the emergence of card-not-present scams, using the security codes on the back of cards, sees fraudsters find ways around such protections. The development of dual-factor authentication makes online card fraud harder, even if a card holder has had their personal account details stolen (online or otherwise). The capacity to breach forms of encryption that most people have on their computers (usernames and passwords) by means of malware is greater than the capacity to socially engineer fraud in the way Cross described. However, the capacity to breach two devices at the same time (the potential victim's computer and their mobile phone) requires far more resources. With encryption on both devices, and with encrypted keys sent to one device to be then used to give authorization on a second device, creates a very narrow time window, while at the same time requiring an even wider (double) window in terms of what a fraudster has to breach in order to access a victim's account. Where social engineering aims to persuade a victim to hand over information that would unlock (de-encrypt – metaphorically or literally) access to personal data, new forms of authentication are designed to make even such disclosure insufficient as digital keys also require physical access to the victims' devices.

4. Technical and social skills

Confidence tricksters in the past combined social skills with technical artefacts. Today's digital fraudster typically combines elements of both in what is now referred to as 'social engineering'. The combination of technical and social skills still applies even if the specifics have changed.

Aunshul Rege (2009) outlines the range and character of perhaps the most intensive form of socially engineered fraud online – that of romance fraud, or sweetheart swindles. While classified ads and lonely hearts columns in newspapers have, for a long time, been a route to conduct such attempts to defraud strangers, and everyday life has always had its share of such deceptions, the internet represents a significant force multiplier. Rege gives four reasons for this: the remote and round-the-clock level of access provided by online media; greater anonymity (concealment); new forms of interaction (incitement); and the speed of matching by which fraudsters can both appear and disappear (concealment). The scale of online romance fraud is hard to quantify as many incidents may never be disclosed due to shame and even ignorance on the part of the victim (Whitty and Buchanan, 2016). However, Rege identifies that the most prolific of these schemes (whether organized by groups or carried out by single individuals) cross international boundaries and, hence, make policing them all the more difficult – not least because jurisdictions and laws differ, and police forces rarely prioritize what is often discounted as a product of the victim's own technical and moral failings.

What is particularly significant about online romance fraud is the combination of sophisticated emotional manipulation and relatively low-level technical skills, at least at the level of interaction (usually via direct email or through posting false or misleading profiles on dating websites). Some sophistication may be required to acquire personal details of individuals who can then be posted up as the false prospective partner, and identity theft of this type may extend to stealing that person's other personal data for the purpose of applying and paying to join dating sites. However, even this level of technical skill is not required if the fraudster simply posts false information that is not based on any real person, or even posts genuine information about themselves, when they have no intention of following up on any promises made. Fraud victims are manipulated, usually on an escalating scale, such that initial romantic interactions and claims of affection are only later followed by pleas for money based on either the need to travel (to visit the victim) or to pay for coping with tragic events that the fraudster claims have occurred to them. Rege refers to this escalation as 'the cycle of lures'. Whether through bulk emails, or by posting on dating sites, fraudsters are able to lure a self-selecting group of prospective victims, already manifesting the need to be liked and so prone to being manipulated by someone willing to tell them what they want to be told (that they are liked or loved). Rege notes the common features of successful scammers: patience (the long game); relatively low technical skills; knowledge of routines by which to draw victims in (social engineering: telling people what they want to hear, or manipulating them with emotionally charged claims if they do not pay – for example, 'you don't have any feelings for me' or 'I thought we had a real relationship here'); and a series of common neutralization techniques by which perpetrators claim they are simply giving people what they want (positive justification) or deserve (negative justification). It is also interesting that scammers and their victims are relatively equal in terms of gender composition, with comparable numbers of men and women in both roles. Romance fraud can blend into extortion as internet interactions are usually recorded (whether in textual or visual form) and, if fraud does not work, perpetrators may turn to threats of revealing compromising material (which may be sexually explicit images, but may also include other kinds of embarrassing or compromising information).

As was noted by Cross (Section 3), in relation to advanced fee fraud, there is also a strong tendency to blame victims for becoming victims of romance scams. As with extortion-based scams, romance scams play on intensified emotion – even as the former plays on the disruptive power of fear, whereas the latter draws on desire in order to disrupt the victim's capacity to rationally evaluate their situation. Beyond the victim's rational capacity to evaluate their situation online, Bradford Reyns (2013) used 2008 and 2009 data from the British Crime Survey (now the Crime Survey for England and Wales) to examine predictors of victimhood in relation to online fraud. Reyns found that using the internet for the purposes of online banking increased the risk of being the victim of internet fraud by 50 per cent, while using the internet for online shopping increased the risk by another 30 per cent. In the years after Reyns' study, the number of people using one or both of these forms of online financial service has escalated rapidly. According to Statista (2021), 76 per cent of UK adults use online banking, up from 50 per cent when Reyns' published and from 35 per cent when his data was collected. Pew Research (Fox, 2013) found that 51 per cent of citizens in the United States used online banking in 2013, while the figure for 2021 was 74.6 per cent (Business Insider, 2021). The scale of online shopping has risen at an even faster rate and to an even higher overall level. As such, those in the high-risk categories of Reyns' research are now the vast majority of the population. As internet banking and online shopping become routine and increasingly hard to avoid, the idea that the victim is to blame for putting themselves at risk becomes even more erroneous. Fortunately, financial institutions, who have the greater capacity to provide reasonable levels of encryption and surveillance, are required in most cases to take responsibility for losses incurred by victims of online fraud based on the theft of financial details, or the misuse of online accounts by fraudsters. As a result, institutions are incentivized to maximize security, while customers can use online services with a significant degree of security, despite the vulnerabilities that online interactions create. However, financial institutions are not liable for socially engineered scams, such as romance frauds, extortion scams and advanced fee frauds. These social-engineering fraud strategies are, therefore, more of a problem in one sense, but such strategies are also more time consuming, less easily scalable and more readily identified even by relatively inexperienced victims.

5. Fraud, risk and the dark figure of crime online

The hidden extent of a crime like fraud is exacerbated online as remote and concealed access reduces awareness of being a victim. Nonetheless, it is possible to get some idea about the scale of the problem itself, and the scale of the problem of knowing the scale of the problem. It is also possible to reduce online fraud to a significant extent, despite the affordances to make such fraud easier that are offered by the internet. Cross and Blackshaw (2015) outline the official statistics on online fraud for the United Kingdom, the United States and Australia in 2012. While the figures for the UK was around 4.7 per cent of adults with credit or debit cards who had experienced some kind of fraud in the previous 12 months, the figure for Australia was 6.7 per cent (but referred to adults experiencing any – not just card – fraud that was therefore wider than the UK figure). The figure for the United States (which was for 2010) was much higher, at 24 per cent. However, this data referred to households where at least one person had been a victim in that 12-month period (such that this figure would – predictably – be many times higher, as it was comparing the sum of all individuals in a US household with single individuals in the UK). These statistics are not, then, comparing like with like. Even if it were possible to do so, though, identification, reporting and recording levels are always limited, such that a large amount of fraud will go unidentified by victims, unreported to various agencies that may be thought to be the most appropriate for dealing with reports of fraud, and then unrecorded in any aggregate fashion that would then allow a 'total' to be meaningfully calculated. The digital makes small-scale, high-volume frauds and scams much easier than in the real world, and relative to large-scale, single-hit frauds (online or elsewhere). As such, the scale of underreporting by victims due to embarrassment or a lack of awareness (either of having been a victim in the first instance or else of who to contact), and then the scale of subsequent underreporting by organizations for fear of reputational damage, is huge.

It is in relation to the conditions outlined earlier that Cross and Bradshaw suggest that 'reactive' policing is so very ineffective in relation to fraud online. Waiting for victims to report that they are victims may be both too late and too limited, as so few victims actually come forward in time for something to be done, if at all. It is with this situation in mind that Cross and Bradshaw outline the success of one particular scheme that sought to reverse the routinely reactive mode of policing for a more proactive approach to tackling fraud online (in this particular case, advanced payment fraud). Project Sunbird was initially a collaboration with the West Australia Police and West Australian Department of Commerce, but then went on to develop further collaboration with five West African police forces and prosecutors. The project set out to tackle proactively advanced fee frauds originating from West Africa and specifically targeting citizens of Western Australia. The project involved identification of payments being made (and screening out those that did not seem suspicious), then writing to the person sending the money to explain why the police/department of commerce believed the payments were part of a fraud scheme. Following this, the department of commerce, via remittance agencies, blocked ongoing payments. The next stage then involved intelligence gathering by means of interviews with those identified as potential victims. The final stage involved collaboration with West African police forces in targeting and prosecuting fraudsters. The first round of letters saw a two-thirds reduction in payments, and a significant reduction in the amount being sent by those still sending. A second round of letters had an additional and significant impact. The project also increased intelligence, while also increasing the visibility and legitimacy of victims (reducing the stigma and reducing isolation). The partnership approach, between agencies in Australia and between police forces in different countries, also challenges the notion that 'routines' and 'jurisdictions' represent insurmountable obstacles to tackling fraud online. Of course, it is important to note that such proactive policing does run the risk of targeting particular avenues and particular perpetrator stereotypes that may then become self-fulfilling prophecies. If more resources are focused on West African (419) fraud scams, this will make such scam origins more likely to be identified and hence reinforce the view that crime is focused there, while the very act of reducing resources in other areas may also, thereby, increase the belief that other places are not a source of problems.

Levi et al (2017) make the point that a very significant proportion of online fraud has at least some part of the process operating within the nation in which the victim resides. While some scams can persuade victims to send money abroad, this is always going to be an added obstacle to inciting people to pay. While some romance scams are international, geographical distance is a significant barrier, and persuading a victim to pay for (false) travel plans, while potentially lucrative, may also dissuade many potential victims. Levi et al (2017, p 11) note that 97 per cent of online fraud victims in the UK believed the perpetrator was operating in the UK. Not all of them were, but phone and text messages from abroad, and sending money to accounts offshore, were disincentives. As such, the claim that jurisdictional issues are a fundamental barrier to policing fraud online is not correct (and, as Cross and Bradshaw have pointed out, it can be overcome anyway). The real problem, Levi et al suggest, is one of 'problem ownership', the sense of 'market failure' in guardianship. Different agencies – formal police and other regulative and service-providing enterprises – may assume that the responsibility lies with one or other agency, not them. The key, then, is to develop collaborations. The combination of collaborative protection initiatives and preparation strategies designed to better inform end-users of various digital network services is unable to eliminate online fraud. However, it can (and does) have the effect of raising the costs of committing such acts (in terms of both upfront organization of scams and the cost with regard to prospective punishment), and reduces the overall rewards that can be generated from such acts.

In conclusion

It is in the nature of fraud that the victim is incited to act against their own best interests by means of some kind of deception. The capacity to access greater numbers of prospective victims by means of digital networks, and the greater capacity to engage in forms of concealment online (relative to being able to impersonate others in real-world situations), does radically increase the scope for fraud by means of online methods. The very nature of digital networks, and especially their use to engage in commercial transactions, has required the development of greater means of self-alienation. By such means, identity (in terms of means of identification) and identity (in the sense of the integrity of the person or self) can be separated, such that the former can be accessed remotely, or used to access other things and people at a distance. Such detachment means both that victims can suffer more readily from 'identity theft' and that perpetrators of fraud online can use such identities to conceal themselves. However, levels of fraud online are not out of control. The capacity of service providers to offer relatively secure encryption is high (even if users are not always so aware and may undo such measures by giving away keys to what are otherwise relatively secure forms of encryption). Also, the capacity of law enforcement to cooperate both with service providers and across jurisdictions is itself afforded by digital network methods too. The scale of concealment and evasion afforded to fraudsters is not absolute, while the scale of returns can be reduced by the combination of user education, proactive policing and various levels of geographical and interorganizational cooperation.

9

Sharing Software, Music and Visual Content

Key questions

1. How is the rising significance of intangibles (such that increasingly non-rivalrous goods can be digitally copied/shared at no marginal cost by end-users) changing markets to piracy and property rights monopolies from sales to advertising vehicles?
2. Does copyright infringing sharing online kill incentive or is it a wellspring for creativity?
3. How has digital distribution transformed the music industry?
4. In an age of livestreaming (where copyright-infringing services compete with copyright respecting services), how can film and television industries survive and even thrive?
5. In the absence of a 'live' product for free sharing of recorded content to help sell, how is it that software and gaming industries have adapted best to free copyright-infringing alternatives?

Links to affordances

Not everyone has internet access, nor know all its affordances. Nevertheless, digital networks make access to copyright-infringing copies of intangible goods instantly available to billions, everywhere, all the time. Attempts to limit access by means of prosecution and blocking have failed due to successfully distributed evasion even when encrypted concealment fails. This is not due to technical inevitability but simply because commercial providers, in seeking to sell content, make it available to others who have an incentive to decrypt (breaking concealment) and to maximize evasion (via torrents, streaming and 'cloud' distribution). Free content creates a radical incentive (incitement) to infringe copyright. Creative industries that best adapt are those that have created services that users are willing to pay for, given that they can no longer be compelled to do so.

Synopsis

How has intellectual property (IP) become so significant in recent years, and to what do new threats to IP represent a challenge? Has the movement towards distributed network sharing thwarted regulation or merely adapted to it? The 1984 Sony ruling in the United States declared that a 'record' button on a video 'recorder' was not unlawful even when its primary utility was to enable copyright-infringing 'recording' of television programmes. The next 30 years saw this principle fought over again and again, and with diverse consequences. Does it make any sense to legislate against objects rather than their uses? The very notion of intellectual property is built upon the idea that creativity resides in objects rather than in action. However, online sharing undermines property rights, increasing rewards to creative actors when people stop paying for recordings and consequently pay more for live events.

Downloading films and, in particular, the free livestreaming of sporting events are said to undermine commercial sponsorship of visual culture. Is this true, and if so, what are the consequences and the alternatives? Cinema attendance/home cinema subscription is up, as is the price of live sports and other subscription television channels despite the ready availability of free streaming alternatives. The most directly digital of all content producers, the software industries, are most vulnerable to online sharing as a threat to profits. However, this sector has adapted better to a sharing economy than traditional media industries. Sharing stimulates development in an industry best able to keep up with alternatives. Sharing has also been key to the development of the network society in the first place. Adaptation to the digital affordances of access, evasion (by distributed networking), concealment or transparency (encryption/revelation) and incitement (by free global promotion/publicity) is key. Industries that have embraced/ developed these affordances have thrived, relative to industries that seek to survive by entrenching outdated industrial models of ownership in a new informational mode of development.

Chapter sections

1. Intangible goods – in particular digital content, but also patent and design information that can now easily circulate online – have become increasingly valuable, and therefore prone to infringement (unauthorized copying). Value and hence desire to infringe is the result of stronger laws, not the cause of them. Regulation of property alongside deregulation of production and wages is an asymmetry promoted by particular lobbies.

The gap between deregulated production costs and regulated monopoly prices incites infringement. In different ways free-sharing networks extend access, while pirate capitalists recreate markets (Green, 2012; David and Halbert, 2015, 2017).

2. The rise of online sharing parallels the emergence of global network capitalism and repurposes many of the affordances offered by networks created to further private corporate expansion and evasion of state regulation (see David, 2017a, 2017b and 2019b).

3. Far from creating a parallel to Rachel Carson's 'silent spring', the rise of peer-to-peer file sharing, torrents and streaming has not killed the music industry, but, rather, afforded a shift from an emphasis on recording to one focused on live performance – to the benefit of audiences and performers alike (see Krueger and Connolly, 2006; David, 2010).

4. The new mixed economy of sharing, pay-to-view and subscription payment in film and television sees content made available for free, even as more people are willing to pay. Free alternatives have forced commercial providers to offer more attractive packages to potential paying customers; but free sharing also aids the biggest commercial players, as free content limits the scope for new and/or smaller commercial operators to enter the field and to challenge the big monopolies (see David and Millward, 2012; David, Kirton and Millward, 2017).

5. Gaming and other software companies have shown that the best way to adapt to free-sharing alternatives is to offer a better product, not to hide behind IP monopolies (see Lastowka, 2015; Lee, 2015). Policing everyone, everywhere, all the time may seem an attractive option for old media companies, but adaptation and attraction mean that the software industries now dwarf their old media competitors.

1. The rise of intangibles

The increased significance of intangible goods today – in particular, digital content (music, games, software, film, recorded television and live sports television), but also patent and design information that can now easily circulate online (such as via crypto markets for counterfeit/fake goods) – is the result of stronger laws, not the cause of them. Regulation of such property in a global network economy of deregulated production and wages is not an inevitable asymmetry, but the gap between deregulated production costs and regulated monopoly prices means high profits and, therefore, the incentive to infringe – whether freely beyond markets or in the form of pirate capitalism that recreates markets.

Intangibles refer to immaterial goods protected in law by intellectual property rights (IPRs). Guilds and charters, as well as royal patents, existed before the rise of modern capitalism. However, in capitalist society, where property ownership is the primary source of power, the protection of creative works, inventions, designs, marks and a range of other related immaterial goods as 'property' was formalized in a range of new laws. The statute of Anne in 1709 created the first copyright law (covering Great Britain and its then colonies). This statute granted a 14-year protection to literary works, and did not include maps, music or visual works. Fourteen years was selected because it was the time given in medieval guilds to allow for royal patent holders to train up two generations of apprentices when apprenticeships lasted seven years. Later extensions of copyright law across the world saw duration extend, as did the range of content covered, to include, for example, music. Only in the 19th century did UK law place intellectual property (IP) with the author of the work, in distinction from copyright being held by the publisher. Again, it was only in the 19th century that copyright was extended in duration to carry over beyond 14 years. The 19th century saw a set number of years' protection even after the death of the author (David, 2006). Beyond creative works (where expression is protected, not an abstract idea or its concrete manifestation – such as a book), modern patent laws, as well as those covering trademarks and designs, agricultural plant and seed rights and geographical indication (protecting names in relation to locations of original production), have developed. In relation to digital networks, it has been copyrighted goods that have been at the cutting edge of legal infringement, as the sharing of creative expressions (music, film, text, games and live-sports-event broadcasts) are all covered by copyright, and can be most readily copied and circulated digitally. The rise of 3D printers may make patented, trademarked and design-right-protected goods available to digitally copy en-masse in the future. However, at present, physical goods cannot be 'downloaded' or 'streamed' online in the way that music and film can. Nonetheless, online auctions and other sales sites do make trade in IPR-infringing copies of physical goods with design-, patent- or trademark-based elements much easier than in the past (David and Halbert, 2017).

The significance of IPRs, and hence the significance of any increase in the ability to infringe such immaterial property rights, has shifted over time. Over the course of the 19th century, various emerging industrial capitalist societies adopted more or less stringent enforcement over such rights (Johns, 2009). Such differences reflected divergent judgments over the relative merits of protecting creativity in order to incentivize future creation by rewarding it, and promoting competition by avoiding long monopoly controls such that a rights holder would have when nobody else could use their creation/invention without permission and payment. Legal regimes differed in different countries at different times, with more powerful countries tending to favour greater protection and emerging economies preferring the benefits of unprotected access. During the Cold War, Western powers took a relatively relaxed attitude to IPRs, as this enabled more rapid development of allies (in East Asia, in particular), and as the Soviet Union and (at that time) China did not accept such rights in any case. As the Cold War came to an end, the United States became increasingly concerned that foreign economies were developing faster than theirs was, and that, as the world's leading power in both creative works (music, film and computing) and engineering and science, the US was not benefiting from such immaterial property as they felt was theirs. In the 1990s and beyond, the United States became the key driver in creating and enforcing a new global IPR regime, manifested most powerfully by the 1994/5 TRIPS Treaty (Trade-Related Aspects of Intellectual Property Agreement), which requires all signatories to sign into domestic law the full protection of foreign IPRs. Non-signatories are subject to trade sanctions (May and Sell, 2005).

TRIPS has been followed by a raft of bi- and multi-lateral treaties signed since 1995. These have been created under the auspices of the World Trade Organization (WTO) (whose first act was the TRIPS Treaty); the World Intellectual Property Organization (WIPO); and by various regional economic blocs in Europe, Asia, Africa and the Americas. These treaties have created a significant asymmetry within what has, since the abolition of the Soviet Union, become a globalized capitalist market economy. On the one hand, the WTO has promoted global market deregulation, such that increased competition has reduced the cost of labour and, thereby, of manufacturing and transport of physical goods. On the other hand, the WTO's TRIPS treaty has increased and harmonized IPRs, such that those goods protected by copyright or patents have monopoly protection for longer durations and over wider (global) jurisdiction. Whereas copyright was originally for only 14 years, today, copyright lasts for the full life of the author and for an additional 70 years after the author's death. Patents have also been extended to last for 20 years, but various forms of 'evergreening' allow lucrative products to stay covered for additional spans (David and Halbert, 2017). The distinction between discoveries (that cannot be IPR protected) and inventions (which can) has been pushed back to allow increasing patenting of living things and processes. Also, the distinction between idea and expression has also been pushed back to allow increased protection on the grounds of 'look and feel', such that ever more generic forms of expression come under the protection of copyright. The cost of making things has fallen through intensified market competition, while the price of IPR-protected goods (such as the very DVDs, books and medicines whose manufacture is now outsourced to the lowest cost location) can be kept high due to monopoly protection (David and Halbert, 2015).

The rise of digital networks makes it possible to copy and circulate intangibles at almost zero marginal cost (Rifkin, 2014). Such loss of scarcity challenges market value. People tend to discount even the use value of things with no exchange value/price (Green, 2012), setting a price on things that can be accessed for nothing becomes challenging. The ability to make copies to sell – such as in counterfeit goods, where monopoly prices far exceed production costs – creates a significant incentive for 'pirate capitalism', where criminal markets seek to profit from the gap between market-level costs of production and monopoly prices. Where counterfeit products fake a product's trademarks, patent-infringing products use a protected invention (such as the specific chemical formula of a drug). Counterfeit (falsely labelled) medicines (where intangible value lies in the branding) may or may not contain the patented – but still tangible – chemistry that the original does, in fact, use. A copied music album or film may or may not seek to counterfeit its packaging, but is infringing copyright if the content is being copied without permission. When a free music or film site is offering access to copyrighted material at no charge to the end-user, that site's ability to generate revenue from advertising links means it can be seen as a form of 'pirate capitalism'. Rights holders argue that such infringement is a form of 'theft'. The term 'theft' may, though, seem an odd term to describe 'copying'. In copying, the original is not removed in the way theft might at first be imagined. However, theft may refer to many kinds of loss (see Green, 2012, for an account of 13 different ways to 'steal' a bicycle). Yet, the claim that giving access to others to make copies of IPR-protected things removes the incentive to create is contested (Litman, 1991), as creative artists rarely even know what is and is not protected by copyright (and what they think they know, and might thereby be motivated by, is most often incorrect). Overprotection of rights holders (usually corporate actors rather than artists) is rarely the best way to reward (and thereby incentivize) creativity (Lessig, 2002, 2004). Free sharing of creative content is called theft by some, but might equally well be described as publicity, as long as this publicity can then generate sales of something else. A gift economy can be more rewarding than one based on intangible property, as long as there is something people are willing to pay for once they have accessed intangible contents (Currah, 2007). If intangibles are only the menu, what then is the meal (Vaidhyanathan, 2003)?

2. Global network capitalism and sharing

Online sharing developed alongside the development of global network capitalism; repurposing many affordances provided by digital technologies created for corporate expansion beyond state regulation. File sharing, in its many forms, is where networked computer users make available files for others to download or stream. When such files contain copyrighted content, then the act of sharing such files may infringe the IPRs of the rights holder of that piece of music, film, game or other material. Whether the copy is said to have been made by the person uploading, the person downloading it or the intermediary service through which the two others access one another has been central to various legal strategies of regulation and attempts to conceal infringement and evade regulation. Whether a person making a copy is thereby incited not to purchase the work is again open to dispute. Sales of records did fall when access to free copies increased, but this relationship is not a direct one and sales of concert tickets also increased when access to free recordings rose.

File sharing over digital networks can be carried out by simple one-to-one exchange with known others, or over simple net bulletin boards where content can be shared between existing members of a group. However, file sharing became seen as a significant challenge to copyright holders (initially, in the field of music) when three technical elements came together: first, formats to compress content, which enabled files to be shared at speed over limited bandwidth channels; second, mass public access to digital networks over which such content could then be sent; and finally, search-and-connect services that would allow uploaders and downloaders to find one another. Compression formats (MP1, 2 and 4) were developed by commercial television and record companies for the purpose of enabling the production and mastering of recorded music in multiple locations (which then makes it much easier to pay tax, if at all, where rates are lowest), and for the development of cable and satellite television companies. Earlier still, the development of the compact disc (CD) and the digital versatile, or video, disc (DVD) had radically reduced the cost of content storage (relative to records and tapes). That the CD was never encrypted, so as to increase sales (David, 2019a), would later come back to haunt record companies (Sandell, 2007). The internet and the World Wide Web were developed first by government and scientific networks, and only later (Web 2.0 onwards) were developed further, by and for commercial purposes, but this later stage brought networked computers into billions of homes. It was only really the final stage, the development of dedicated sharing platforms, that meant uploaders and downloaders were easily able to make available and access each other's content for the purpose of making free copies.

The history of file sharing online did start before 1999 (with these bulletin boards; see Carter and Rogers, 2014), but it was in 1999 that the Napster file-sharing service was launched as a more user-friendly version of that third step beyond storage and transmission, that is, processing (David, 2016). Napster's central server channelled information to and from those, making files available on their networked computers and those seeking to download copies. This led to the accusation that Napster was itself engaged in 'piracy', the illegal production of unauthorized copies of copyrighted material. It should be noted that 'piracy' is distinct from 'bootlegging'. Bootlegging is where someone makes unauthorized copies of outtakes and of unreleased recordings, and may sometimes extend to making rough copies of recorded content, but with these lacking in the counterfeiting features of a legal recording's packaging. Bootlegging (Marshall, 2005) was not formally approved of by record companies, but as it tended to provide supplementary materials that were not in direct competition with official releases, it was not generally taken to be a serious threat. 'Piracy' on the other hand, in offering access to exactly the same content as official releases (albeit in the form of digital downloads, without the packaging that would render it counterfeiting), did represent a direct challenge to the business model of record companies (and all producers of copyrightable content once the bandwidth sufficient to share it became available – film requiring more than music, and livestreams requiring more than film). Napster operated for only two years, and was forced to close when sued by the Recording Industry Association of America. Napster claimed it was not making any copies, only offering a service that users could do with what they wished (including the sharing of content that was lawful to share). In 1984, the US courts had rejected the Motion Picture Association of America's (MPAA) claim against Sony, that Sony's newly invented video recorders should be banned and the company held liable for users' acts of recording copyrighted content with them. The MPAA claimed that having a 'record' button on such machines was an incitement to record television programmes, almost all of which were copyright protected. The court judgment was that 'dual use' (when something has a legal use, as well as the potential to be used in the conduct of a crime) meant that it was wrong to ban something simply because it had the potential to be used in an unlawful manner (which would require almost every object to be prohibited). Napster's claim that its software was just like Sony's video recorder was rejected on the grounds that Napster's central server meant all files passed through their hands, and so, when such content was infringing content, the service provider was in effect 'handling stolen goods' (David, 2010).

Napster shut down, but not before a series of new file-sharing services emerged with what were fully peer-to-peer forms of sharing, such that users downloaded the software from the service provider and then shared content between themselves without the need for content to be 'trafficked' through the service provider's own computers. This evaded the legal liability of the service provider (in line with the 1984 Sony ruling), but legal efforts to target them on the grounds that they actively promoted illegal sharing then arose. Service providers then actively avoided making any such claims in promoting their services. Sites such as Grokster, Morpheus and Kazaa then arose with distributed peer-to-peer sharing services. This saw legal moves by record companies to target those it was claimed had uploaded content. A number of uploaders were prosecuted and threatened with huge fines estimated on the basis of the retail price of all the copies that had been made from their uploaded file/s (Zentner, 2006). Subsequent research suggests that downloads and free streaming services cannot be said to directly 'cause' reduced sales (Aguiar and

Martens, 2016), even as those that use new legal streaming services are also those most likely to use illegal downloading services as well (Borja, Dieringer and Daw, 2015). Sameer Hinduja (2012) found that file sharers have a little less self-control when it comes to music than non-file-sharers, which rather suggests they love music more. Prosecutions for uploading music files were rarely successful in recouping revenues. However, they did cause file sharers to migrate to more concealing torrent-based services, where any particular download is made up of a series of elements from multiple uploads, such that no one upload can be said to have been copied to a significant degree in the making of any one particular download. Torrent services (most famously The Pirate Bay, or TPB, in Sweden; see Schwartz, 2014) were targeted on the grounds that they actively promoted the use of their service for the infringement of copyright. Jail time was given to some service providers, but services like TPB simply evaded such actions by relocating their servers to jurisdictions that did not enforce such take-down policies (David, 2010). ISPs have been required to block such services in some countries, but VPNs can get around such blockages (Brown, 2015). While torrent services concealed users (this affordance preventing uploader prosecution), relocating services evaded regulation (affording service providers protection from prosecution). By the time new laws had emerged to track and block users, servers had moved onto new formats of access, evasion and/or concealment. When viewers view content on the cloud without making a 'copy' has a crime taken place? If so, where, and should such sharing without copying be a crime at all (Bernat and Makin, 2014)? When is it, and should it even be, a crime to listen to someone else's copy of a record?

Perhaps most significant of all, in the cat-and-mouse games of free sharing and legal attempts to prevent it, has been the incorporation of downloading into legal form in recent years (Casadesus-Masanell and Hervas-Drane, 2010). While, in the 1990s, record companies had resisted any attempt to provide a single repository for the sale of online music, the impact of Napster, and the services that immediately followed its closure in 2001, forced record companies into negotiating a deal with Apple for the creation of the pay-to-download iTunes service in 2004. Without the pressure of free alternatives (which were far more efficient and user friendly than what record companies were willing to offer – with multiple company-specific platforms and encrypted formats), it is unlikely that iTunes would have come into being. It is also important to note that iTunes abandoned its Fair Play encryption software in 2007/8, despite record company objections, as it could not otherwise compete with free-sharing alternatives. While iTunes was a pay-to-download alternative to free sharing, the development of subsequent streaming services, such as Spotify, where most users gain access to free content in exchange for advertising (although some pay a subscription to avoid advertisements) has now come to replicate free file-sharing services almost in full (Da Rimini, 2013). Spotify is a legal adaptation that arose in the face of the unlawful application of network affordances that enabled infringing streaming services in the late 2000s, just as iTunes was a legal adaptation in the face of earlier copyright infringing downloading services made possible a decade earlier. Advertising revenues that once went to the infringing service providers now gets largely paid over to rights holders. Sadly, because rights holders are most often record companies whose contractual arrangements with artists are highly unfavourable to the latter, the amount of income that then returns to artists from services like Spotify is usually negligible (Marshall, 2015). In terms of how artists do (or most often do not) benefit from their recorded works, this is 'business as usual', but free sharing may have, in fact, created something else that is much better (Royle, 2013) – in inciting a rise in live performance revenues, as the next section will discuss.

3. The rise of live

Far from Rachel Carson's 'silent spring', peer-to-peer file sharing, torrents and streaming has not ended musical creativity. Rather, a shift has occurred from recording to live performance. Audiences and performers have benefited from this. In 2008, the BPI (formerly the British Phonographic Industries) released a report (BPI, 2008). In it, they claimed the free sharing of music on the internet would do to music what pesticides had done to bird populations in the United States and elsewhere in the middle years of the 20th century, as documented in Rachel Carson's (1962) book *Silent Spring*. Indiscriminate use of technology would kill off music, the report suggested. Nothing, however, could be further from the truth (David, 2010).

Music piracy goes back to the piracy of sheet music in earlier centuries (Alexander, 2007), and then of unauthorized production of records and tapes. Digitization in recent decades has been a part of a wider transformation of capitalism and the consumption of immaterial things, for which music has been at the cutting edge (Hesmondhalgh and Meier, 2018). Prior to the advent of free sharing online, the digitization of music had already happened. This took place (as mentioned in Section 2) in 1982 with the advent of the compact disc (CD). Sony and Phillips developed the CD in collaboration. Sony was at that time mainly a manufacturer of electrical goods and only subsequently became a record company (now the world's largest). The roll out of the CD saw records and tapes gradually decline, to be replaced by the new format. Initially, CDs were considerably more expensive than either records or tapes but were said to be more durable and of better quality. Those who transferred over to the new format were also then required to reformat their existing record collections. The cost of making CDs was much lower than was the case for records or tapes, breakages were less, and digital content could be more easily transferred to allow instant mass or niche production anywhere in the world without the need for traditional (high, fixed, cost) vinyl presses. The result was the largest profit storm in the history of the recording industry as reformatting increased sales, prices were higher, and costs were reduced. Record companies bought up the previously separate field of music publishers. This allowed them to control more of an artist's potential earnings (the 360° contract – see Marshall, 2013; as well as Curien and Moreau, 2009), and major labels had money enough to buy out independent labels if these had signed new talent that majors could then profit from rather than compete with. Record companies were also themselves subject to and drivers of horizontal integration within increasingly globalized, cross-media, transnational corporations. The CD had been made available in an unencrypted form to encourage its take up as an industry standard. In this regard, the CD's open format was very successful and very profitable. However, with the advent of home CD burners in the mid-1990s and, more significantly, with the advent of mass-access file-sharing services after 1999, the feast very soon turned to famine.

The digital profit storm for major music labels from 1982 to 1999 was followed by a crash in sales revenues, which labels claimed was the direct result of free file-sharing alternatives. The claim made was that what hurt record companies also hurt artists. This rather assumes that business as usual was good for artists. It was not. Steve Albini's online article 'The Problem with Music' (1993) documents how an emerging artist may succeed in getting the fabled million-dollar record deal, and then outlines what such success really means, concluding with the line: 'Some of your friends are probably already this fucked' (1993, np). Having signed a contract, the promised money is, in fact, largely 'in kind', to be 'spent' on the record company's own services to the artist (production, mastering, management, legal services and publicity), while most of the rest will be spent on other costs incurred in promoting sales of the work. The artist may, if they are lucky, be given some limited amount of money to buy food and pay rent. Albini goes on to show that while almost all upfront money outlined in a contract will be spent by the record company, and will not go into the artist's pocket, the artist is still then required to repay all of this money. This might seem fair as the costs incurred are real, even if often exaggerated, but there is a catch. Artists are required to repay this money, not from the sum of net sales, but rather from their royalties, which will be between 5 and 15 per cent of net sales. The traditional record contract, thereby, requires artists to pay for the greater part of the expense invested in their recorded works, from only a tiny percentage of that work's sale value. While many recordings do not repay the investment made in them by record labels, it is also the case that recouping (which is the repayment of costs from royalties) rarely ever happens even when overall net sales do repay record company spending. Record companies can reasonably claim that they have to make profits on successes to cover the cost of works that do not cover their own costs. However, if both successful and unsuccessful artists end up indebted to their labels, the overall business model cannot realistically be said to create financial incentives. Brian Holmes (2003) found that only 1–2 per cent of musicians with record deals make earnings equivalent to minimum wage from such contracts. The difference between recouping and repaying means that all but a tiny handful of artists end up owing their record companies money even when their records do make large profits. A record that sells 1 million copies at (for example) £10 per record will net the record company £6 per copy (as part of the final price will go to the retailer and transporter), meaning, in total, a return of £6 million. If the artist is lucky enough to have been given a 15 per cent royalty, this would entitle them to 15 per cent of that £6 million, which is £900,000. Repaying (recouping) the £1 million advance means the artist now owes the record company £100,000, despite the record company already taking £5.1 million as the remaining 85 per cent of the overall net sales revenue. Recall, this is for an artist lucky enough to sell a million records! Failing to recoup means the record company can now take from the artist's publishing rights (earnings from other artists covering or otherwise using an artist's work) if the record company owns the publishers – which they now largely do. The record company can also take from artists' live performances/ merchandizing sales if the artist has signed (or can be made to sign, due to their indebtedness) a 360-degree contract. Unable to get out of their contract due to their ongoing failure to recoup, the artist is likely to be offered even worse royalties on future works. Courtney Love (2000) suggests that this situation means artists are better off giving their recorded work away for nothing, in the hope that such free publicity will increase their earnings from live performances (see also Jeremy Gilbert, 2012). This is what, in fact, has happened in the two decades since Love's piece was published. Opportunity costs refer to the cost incurred in doing one thing in terms of all the other things you cannot then do instead: if I spend £10 on a recording, I then have £10 less to spend on other things. Alan Krueger and Marie Connolly (2006) examined the relationship between the fall in sales of recorded music in the years after Napster (and subsequent peer-to-peer sharing networks) arose, and the rise in ticket sales and ticket prices for live music events. The relationship is very clear. Less money spent on recordings caused a significant increase in the number of tickets sold and the prices music fans were willing (and able) to pay. Free circulation of music may have also contributed to an increased valuation of music relative to other spending alternatives over and above simply the transfer of spending from recordings to live performance (Derwenter, Haucup and Wenzel, 2012). Given that artists do so very badly from the traditional record contract, their earnings from live performances were always greater for all but a very tiny number of very famous artists. If the decline in record sales increased overall live performance revenues, as Love suggests, it is better for artists to give their recordings away in the form of publicity, as this 'advertising' increases their incomes from live performance (David, 2013). Robert Morris and George Higgins (2010) suggest that file sharers learn from each other the idea that 'piracy' is acceptable. Such a social learning theory approach might also explain why many musicians also approve such IPR-infringing practices among their fans. For record labels that have adapted to downloading and streaming, the digital has not meant death, but only adaptation (Rogers, 2013). Nevertheless, the potentially infinite reproducibility of digital content does present an ongoing challenge to notions of private property based on the idea that scarcity is inevitable and that, therefore, things must always have a price so as to enable selective allocation (Arvanitakis and Fredriksson, 2016; David, 2019b). The expansion of live performance is itself precarious, as witnessed during the COVID-19 epidemic. COVID-19 saw the in-person, live music economy shut down in 2020–2021, causing a huge loss of earnings. Many artists found new ways of performing and connecting with paying fans via digital networks, but, at the time of writing, the live music economy is still to fully recover. Streaming forms of digital access have facilitated payment and infringement, inciting some to pay to view content online, or to buy real-world concert tickets down the line. This was also true of earlier infringing download services. Even if some/most viewer/listeners do not pay, those that do reward artists at a greater rate than did/do returns from record sales. Infringement is still the best way for artists to get paid.

4. Visual media

Early livestreaming services were multipurpose, without being specifically dedicated to hosting live sports broadcasts. Early services like Justin.tv allowed users to create their own channels and then to broadcast whatever they liked. Some users choose to livestream copyright-protected sporting fixtures without permission, leading to attempts to prosecute the service providers. However, channels like Justin.tv, which did not specifically promote themselves in terms of free access to sports content, could evade liability as long as they removed infringing content after it had been brought to their attention. If it took a matter of hours to be notified and to check the content being streamed before taking it down, then the fixtures being streamed would most likely already have been finished. This temporal evasion did not last. Soon service providers were required to enact automated take-downs in order to comply with the 'dumb pipe' exemption from liability, and this saw streams shut down mid-game. However, with multiple other streams opening up, users could easily reconnect. Still, such disruption reduced the spectator experience. Soon, alternative evasion strategies arose, with offshore streaming channels being located in less restrictive jurisdictions (such as Mexico, in the case of First Row Sports). Courts in national jurisdictions were able to require local ISPs to block access to specified channels, but other channels have arisen, while dedicated streamers can readily turn to VPNs to bypass such blockages.

Free-sharing, pay-to-view and subscription access to film and television creates a situation where overall payments increase while free access also increases. Sharing services compel commercial providers to offer better services to potential customers; but free access also helps dominant commercial providers. This is because sharing channels reduce the potential for commercial rivals to enter the field and to challenge existing first movers.

The digital revolution in music, with the advent of the CD in 1982, preceded a second digital revolution, with the advent of Napster's free file-sharing service in 1999. Similarly, a 17-year gap exists between the development of digital sports broadcasting in the UK in 1992 and the first major signs that free online streaming services were challenging the first round of digital service providers (Birmingham and David, 2011). However, the double digital revolutions in music and in live sports broadcasting did not follow identical pathways (Kirton and David, 2013). The digital profit storm that followed the invention of the CD flowed from the requirement of users to reformat (buy new copies of existing records and tapes), as well as the combination of higher prices to customers and lower costs to manufacturers. The digital profit storm in the development of digital sports media flowed, meanwhile, from enclosure and extension. Digital channels bought television rights for popular national sporting leagues (for example, football, cricket, tennis and rugby) from relatively local, terrestrial, analogue channels that could only afford limited amounts to buy up such rights. These digital channels then sold access to those live events, not only back to those living in the locations where events took place, and who had watched them previously for free via analogue channels, but also to wider audiences across the world via satellite broadcasting. Access to digital channels could be encrypted such that viewers would be required to pay subscriptions or to use pay-per-view in a way that analogue television could not enable, which meant analogue channels could not compete when bidding for rights against digital channels. Scope to sell such access worldwide just made digitization all the more profitable.

The development of digital sports subscriptions created the initial funding for the development of today's more wide-ranging digital television services, but it came at a huge cost to sports fans, who, while having greater choice if they were willing to pay for it, did find themselves having to pay for what were their traditionally free, national league television broadcasts. This became part of what some have seen as an increasingly gentrified version of spectator sport, where once-core fan bases are gradually becoming priced out of both live stadium attendance and (progressively expensive) digital sports broadcasting services (Millward, 2017). Monopoly prices will always create an incentive for pirate capitalists to make black market profits. So in the field of sports broadcasting, once domestic internet bandwidth was sufficient to carry live visual images of any significant coverage, so it was that copyright-infringing livestreaming channels began to emerge that re-routed rights-compliant broadcasts to those who either could not or did not want to pay for them.

Digital networks in the first wave of the commercial digital revolution had allowed traditional fans to be bypassed in favour of wider and more lucrative markets. Now it was the case that streaming services, often accessed by means of the very ISPs who were seeking to sell subscription services, gave these fans a means of bypassing those gatekeepers that, via encrypted broadcasts and escalating season ticket prices, had previously excluded them (David and Millward, 2012). Manuel Castells (1996) once discussed how elite players in networks had successfully marginalized 'support labour' insofar as elites could simply switch from one location to another in the supply of such support labour. So it was that digital networks could also afford some level of counter-power, at least at the level of overcoming encrypted services and circulating content for free (David, Kirton and Millward, 2017).

For as long as it is not possible to eliminate free and copyright-infringing livestreaming, the largest digital sports broadcasters can maintain their market dominance and profitability, precisely because free-sharing alternatives limit scope for legal competitors to gain a paying subscriber base. In the case of English Premier League (EPL) football, the digital broadcaster Sky gained first-mover advantage in the 1990s when it bought up the first rounds of rights to the EPL, which it then used to encourage customers to subscribe to its newly developed satellite television service. In the early 2000s, the UK government felt Sky's monopoly over Premier League football rights was too restrictive and they required that at least some of the bundles of matches be bought (covering periods between three and five years) by other services. New service providers did enter the market, but Sky's first-mover advantage, combined with free infringing streaming services, have meant they failed to thrive. The first of these (Setanta) withdrew in the middle of its three-year rights deal, and was replaced by ESPN (a subsidiary of the Disney Corporation). Setanta withdrew as the subscriber numbers that it had estimated it would achieve when it bid for the rights were not reached, due to free infringing alternatives (David, 2011). ESPN also withdrew for the same reason after two rounds, and was replaced by British Telecom (BT), which has itself struggled to make its sports broadcasting service profitable because it cannot generate enough subscribers. Meanwhile, Sky, which had already recruited 10 million subscribers when it was still a commercial monopoly and before bandwidth and compression allowed free streaming alternatives, remains in its first-mover-advantaged position. That Sky is the largest supplier of broadband internet in the UK means it is the very provider of the resources required to illegally stream the same live sports content it owns the rights to and which it seeks to profit from. That may seem like a contradiction, but given that free alternatives have meant there is only a limited new audience for paying service subscribers for Sky's competition to build upon, keeping the terrain soft around Sky means that those already secure in their established position – such as Sky – cannot ever be successfully challenged (David, Kirton and Millward, 2015). Ma et al (2014) found a similarly contradictory picture when looking at film revenues in relation to pre-release movie piracy. While pre-release piracy does impact on sales, it is still the case that cinema attendances have been maintained in those years since free downloading services have been available. The rise of online subscription-based television streaming services like Netflix and Amazon Prime, in an age where anything broadcast can be re-routed and streamed for free by other means, also illustrates the curious nature of today's mixed economy of visual media.

5. Computer games and software

Software companies adapt to free copiers by offering new and better products, making long term IP monopolies irrelevant. Old media attempts to police everyone, everywhere, all the time are obsolete. Adaptation and attraction, rather that legal controls, allow software industries to dwarf old media competitors today. With relatively recent origins, computer gaming (which only began in 1961, with very simple games in university science laboratories, moving then into arcades and only in the 1980s taking off, in the domain of home computers), today's computer games industry is estimated to generate around $114 billion (US) annually (Statista, 2020). This is around five times the value of global recorded music sales, and only marginally less than total world cinema and home box office revenues. Non-gaming-based computer software sales in 2021 are estimated to be worth at least half a trillion US dollars (Statista, 2022). It should be noted that Apple and Microsoft, the two largest companies in the world by capitalization, are so large because they have become ubiquitous. For many years, Microsoft allowed its software to be pirated outside its core markets to eliminate competition, and Apple's decision to remove its Fair Play encryption software on its iTunes in 2007 was made for the same purpose. Just as a first mover like Sky in the UK (see Section 4) benefits from competition having no purchase in a space where free alternatives exist, so today's market giants have become so large precisely in environments where markets only exist in a very limited sense.

It is also worth recalling that much of today's global digital networking software is based upon shared rather than privately owned software. The internet was developed first by military-research-funded scientists and then by academic researchers (Abbate, 1999); and the maintenance of the internet today, and in particular with respect to the protocols that enable networked computers to communicate with one another, depends upon non-commercial players creating a level playing field (Lee, 2015). The World Wide Web Foundation was created by those non-commercial actors who created the web with the specific purpose of maintaining its non-privately owned nature such that it cannot become the monopoly preserve of profit-driven actors (Berners-Lee, 2000). Apple's creators appropriated earlier software programmes at a time when the Homebrew Computer Club was a collective of amateur hackers, much to the dismay of Bill Gates (1976). The relationship between free and open source software operating systems and commercially produced derivatives is one where the latter has largely parasitized the former (Söderberg, 2008). Software companies complain about piracy, but it is not a serious threat to their business (The Economist, 2012).

The relationship between freely shared content, commercial search engines and social media platforms, again, highlights the paradoxical relationship between what is freely given and services that can then profit from such content. Facebook and Google are the most extreme cases in point, where users gain access to free content on freely accessible platforms where these platforms profit from selling eyeballs to advertisers – not the 'sale' of IP, which is routinely infringed by such services. What services like Napster were legally prohibited from doing (in the past and still), today's social media platforms have rendered legal (with payment to content providers), and, as such, Spotify, Facebook, YouTube, Google and so on have simply turned what were once criminal infringing services into legal forms.

Film and music piracy can incite cinema and live concert attendance, as well as ticket prices – and live sport too, where this has its 'real-world' alternative to what can be streamed. Software, one might imagine, is not like this. Computer games do not have much in the way of any real-world equivalent that the digital version could ever act to 'promote'. As such, should computer games not be the most vulnerable to piracy? Whether or not they 'should' be, the reality is that computer game sales have rocketed precisely at the time when older cultural industries (film and music) have struggled with piracy. How can this be the case? The answer is that the computer software industries have better adapted to keeping their customers in ways that do not rely on (what might, therefore, be seen as) outdated forms of IP protection. When computer games first moved outside of university science laboratories, it was in the form of arcade games, for which there was a lively market in pirated machines. However, it was the rise of the personal home computer (the PC) in the 1980s and onwards that really saw games and games piracy take off, firstly via tapes (when software was contained on tape) and then on the various evolutions of recordable discs (Longshaw, 2011). The solution to piracy via user-generated copies has been the development of both the console market (Depoorter, 2014) and, later, the online multiplayer environment (where only authorized copies or access codes enable use or participation) (Kamenetz, 2013). While any PC owner can access pirated copies of games online, users are persuaded to buy console-compatible (legitimate) copies, or to subscribe to official multiplayer forums (where logging into the official game via a central server makes infringing much harder), on the basis of a better game-playing experience, not on price or necessity (Neiborg, 2016; Hart, 2017). This makes IPR protection largely irrelevant. When popular console games are revised on a near-annual basis (or even more rapidly in some cases), copies of last year's version become next to worthless almost overnight. The idea of copyright lasting for the life of the author plus 70 years becomes something of a joke.

Greg Lastowka (2015) points out the very significant difficulties involved in identifying exactly what counts as 'original' content available to be copyrighted in computer games, as rules of a game cannot be owned. Much graphic content is based upon generic code, so at exactly what point does a particular moving scene become a fixed image (such as an artist might claim was their original work)? The embedding of hyper-realistic graphical (non-game-play) sequences in computer games are there precisely because these can be owned, but such segments often interfere with game play and so reduce the popularity of games. When users generate content themselves within games, objects and sequences of play cannot be easily claimed by game designers either. High-profile legal cases between games companies in dispute over ownership are so very protracted because it is so very hard to get a judge to arbitrate in a space where so much is always already shared and everything that is produced is always changed in every use. Games companies have been far more successful in just getting on with producing content that users want to use; and this is a far better route to commercial success than trying to sue one's competitors, let alone players who do not pay when so many others do (David, 2017a).

In conclusion

Digital compression, bandwidth expansion over digital networks, and search processing software have made it possible to access free IP-infringing copies of cultural works that would have once required physical and commercial manufacture and distribution. Additionally, commercial pressures have led to non-encrypted copies of CDs and DVDs being sold after encryption was routinely breached by hackers of iTunes, undermining commercial attempts to conceal content behind paywalls of various kinds – and this also means access is easy over digital networks. However, early sharing software also made, first, platforms like Napster, and, later, users of more distributed services identifiable and, therefore, liable for infringement. Peer-to-peer services made service providers less liable, but such distribution still allowed uploaders' concealment to be breached and for them to be prosecuted. The development of torrent systems enabled greater concealment of uploaders. Torrent service providers who advertised themselves as enabling free sharing of specifically IP-protected content (direct incitement) were then targeted. Relocating servers enabled a degree of evasion; and while requiring ISPs to block access did conceal services from casual users, this too could be evaded via VPNs. Streaming services enacted an initial temporal evasion strategy (by the time rights holders had reported a channel on such a streaming service and it had been investigated, the event being streamed was already finished). However, since take-down systems have now been largely automated, this strategy has been replaced (as with torrents) by geographical evasion, blocking and countering via proxy networks in another cat-and-mouse game of access, evasion and concealment. However, it is adaptation to the affordances to access, conceal, evade and incite, rather than simply seeking to prosecute such services, that has proven the most significant response to online piracy. Free sharing has incited a rise in live performance audiences and ticket prices, while cinema and subscription television have simply had to raise their quality to retain and expand their markets, and software developers rely on being better than the free alternatives rather than simply seeking to prevent them. Online shadow markets in physical goods and the potential to download formulas to make patented and design-rich physical objects will be the cutting edge of such adaptations going forward.

10

Conclusions

Does the digital really make a difference? Digital access has offered new routes to commit fraud, circulate obscene content, groom, stalk and bully, surveil, and download. Digital encryption conceals identities in relation to terrorist communications, fraud, hate speech, stalking, planting fake news, 'scareware' threats, online obscenity, bullying and trolling. Distributed networks and services afford jurisdictional evasions in relation to obscene and hateful content, fraudulent and extorting communications, peer-to-peer and torrent-based downloading and streaming. Digital communications may incite radicalization, polarization, normalization and/or disinhibition in relation to hate, abuse, dispute and theft. Yet, in relation to all of these outlined, digital networks have alternative affordances that, when applied, may inhibit criminality and protect potential victims (see Table 10.1 for a summing up of what is described).

Digital networks offer four affordances for criminal harm: access, concealment, evasion and incitement. These affordances each intersect with the four domains of criminal harm documented in this book: hate (political and personal); obscenity (adult pornography and violent video games, as well as child pornography); corruptions of citizenship (misinformation and invasion of privacy); and appropriation (fraud, extortion and IP theft). As the preceding chapters of this work have identified, networked technologies are symmetrical in principle (affording greater levels of harm alongside greater levels of protection), but such symmetry in principle is not the same as symmetry in practice. The way digital technologies are used determines outcomes, not some (symmetrical-in-principle) logic of technology itself. In this sense, David Wall's 'transformation test' (see Chapter 1) is not straightforwardly 'passed' or 'failed' in relation to any particular field or affordance. The 'binary' (scope) measure, by which the digital might be said to pass Wall's test, only applies to pure forms of hacking where computers are both tool and target, and this form of hacking is relatively limited. As such, it is the 'quantitative' (scale) and 'qualitative' (severity) measures of Wall's test that are most significant; and in these measures, the relationship between

Table 10.1: Fields and affordances: a symmetry in principle (if not in practice)

		Hate		Obscenity		Corruptions		Appropriation	
		Terrorism	Personal abuse	Pornography	Child abuse	Invasion of privacy	Free speech	Fraud	Piracy
Access	Pro-crime	'The long fuse'	Digital intrusion	Spread of obscene content	Circulation and grooming	Hacking	Disinformation goes viral	High-volume micro-fraud	Zero marginal cost of pirate content
	Anti-crime	Cyber-warfare surveillance	Police surveillance	Following and locating criminals	Identification and evidencing	Digital security technologies	Victims' voices can get heard	Digital verification	Ad-funded free legal content online
Conceal	Pro-crime	Encrypted Comms	Anonymous communication	Encryption curbs police action	Encryption and fake accounts	Hidden intrusions	Disinformation agents can hide	Digital deception	P2P and torrents hide sources
	Anti-crime	Anonymous surveillance	Hidden/disguised guardians	Anonymous surveillance	Honey traps	Encryption keeps some things secret	Anonymous victims speak up	Digital encryption	Hidden trackers
Evade	Pro-crime	Jurisdiction skipping	Inaccessible location	Global relocation	Relocating uploaders	Relocation inhibits prosecution	Channels claim 'dumb pipe' role	Jurisdictional distance	Remote servers
	Anti-crime	Proxy action by other states	Victims find new safe spaces	Increasingly mobile policing	Harmonizing laws and acts	Using foreign apps (Telegram, VPNs)	WikiLeaks etc. servers abroad	Dual-factor authentication	Protected spaces (platforms/consoles)
Incite	Pro-crime	Radicalize/propaganda	Spread conflict and fear	Copy-cat violence	Corrupt viewer and viewed	Promise of free services	Fake news leads to violence	Scams incite self-disclosure	Free infringing content incites use
	Anti-crime	Scrubbing and de-ranking	Online tribes fight back	Mobilizing feminist activism	Promoting cooperation	Whistleblowers fuel disclosure	Citizen journalists encourage protest	New paths inform of risks	Free legal content incites use

symmetry in principle and asymmetry in practice is always contingent. It is also true that symmetry in practice needs to be qualified. Symmetry may still manifest transformation. In such cases, outcomes are 'symmetrical' but technology affords, and may be said to have (in practice) fuelled, polarization of relations – that still remain symmetrical, despite this amplification.

In this chapter, the four fields of crime (hate, obscenity, corruptions of citizenship and appropriation) are summarized in their relation to digital networks; as are the four affordances (access, concealment, evasion and incitement). This is followed by an overall account of the balance of risks online. Does the digital make a difference? Yes, but only when we let it do so.

Hate

The internet, as a distributed network, was initially a US military infrastructure, which became associated with distributed, non-state, terrorist actors, but which has now returned to being seen primarily as a vehicle for state-level information warfare. The relative capacity of states and non-state actors to use networks for encrypted communication and/or propaganda broadcasting favours states once again. The 'logic bomb' has not replaced the 'truck bomb', and early fears of 'the flood' or 'the crash' have been replaced with concern regarding terrorist organization, recruitment and radicalization rather than the direct use (access/long-fuse theory) of the digital as a direct weapon of mass destruction. The suggestion, post 9/11, that 'viral' terrorist cell structures mapped the distributed network age better than states, in terms of concealment and evasion, and that the vertical skyline of New York mirrored the hierarchal rigidities of the defunct Soviet Union relative to new horizontal (digital) network forms, has proven to be an error. Dominant states are better adapted to networks than terrorist groups (Benson, 2014); but states in competition with one another create new forms of symmetry, such as in competition over surveillance and encryption (VPNs, TORs, WhatsApp, Telegram, Zoom, 5G infrastructure and TikTok).

States have introduced new laws against hate crime and hate speech, and even the United States prohibits such forms of harmful language that might incite criminality or compound its harm. The capacity to surveil and dismantle terrorist networks has shifted attention from coordination to radicalization in the form of incitement to carry out 'lone-wolf'-style (low-tech and limited-organization) actions. Beheading videos did have real effects in Iraq/Syria (and in neighbouring countries like Jordan), and likely contributed to radicalizing a very small number of people beyond the region. However, state and mainstream media actors were able to control the circulation of such images and their interpretation, at least in relation to the vast majority of the population outside of the immediate region, and even within it in time.

Today's hackers are more likely to work for states than for non-state terrorist groups, even as such hackers as do work for states sometimes also moonlight (as whistleblowers). As such, the contradictions (or the contested genealogies) of hacking are more likely to play out within the (often-outsourced) networks of an increasingly global, inter-state, security assemblage than within terror cells.

Digital access has seen a shift in the profile of stalking and bullying behaviour towards a more even gender balance, but, where the digital is combined with real-world intrusion, women remain more likely to be victims and male ex-intimates the stalkers. Women continue to experience stalking as more threatening than do male victims. The extent to which digital forms of stalking and bullying offer greater scope for concealment and evasion for perpetrators is mixed, as greater anonymity can be achieved with sufficient technical competence, even as digital mouse droppings make online stalkers and bullies open to tracking in ways that real-world actors are not. For victims and victimizers (very many are both), the digital offers escape from (encryption/safe space), and yet extension (via surveillance) of, targeting.

Finding one's tribe online can create new forms of safe space and community for individuals marginalized in real-world spaces; yet, such online spaces may foster polarization and an increased intolerance to others when different groups do confront one another in online or other spaces. 'Digilantism' and the democratization/normalization of stalking, trolling and other forms of online intrusion/insult illustrate the polarizing potential of online symmetry. Dispute over the meaning of virtual 'intrusion' – 'unwanted' and 'persistent' communications in relation to public profiles and social media platforms designed to evoke and value responses – means that one person's free speech is another person's invasion of privacy and, potentially, trolling/bullying/stalking.

Some male gamers believe online game space belongs to them, even as feminists seek to claim space in such domains while also resisting intrusion into their own spaces by those who would challenge them. The horizontal nature of digital communications aligns with anti-hierarchical resistance to any form of judgement and criticism, even as networks facilitate greater levels of such commentary from those who reject the view that they need any 'authority' or 'expertise' to express their opinions as truth. 'Mobile network youth' are particularly sensitive to online judgement and depend upon networked belonging/exclusion, but are also better able (over time) to adapt and find communities of like-mindedness – even if that may foster greater polarization. At the same time, the language of bullying, stalking and trolling has come to be used by everyone about everyone. Today, few accept that anyone else has the right to criticize them, or that anyone has the right to stop them criticizing others.

Obscenity

Incitement is central to debates over adult pornography and violent video games, but evasion, concealment and access issues compound the 'problem' of incitement if regulation is weakened by such affordances. Obscenity assumes the ability of certain material to incite harm, but the technical separation of sex/violent simulation from direct 'consequences' (babies/victims) can be a positive or a negative thing. The 'murder box's' 'zone of exception' releases frustrations for some, but increases/normalizes hostilities in others.

Laws have been relaxed in relation to obscene content, but new laws regarding extreme content have been passed and some degree of harmonization between states has been developed in relation to global network distribution. However, restriction in itself may act to incite interest and desire, so the relationship between control and what is being controlled is not a simple one. McGlynn and others believe increased access to unregulated pornography fuels sexual violence and misogyny, even as McCormack and Wignall suggest it has increased tolerance of diversity, and McKee suggests such increasing access correlates with increased gender equality over recent generations.

As consent comes to predominate as the primary ethical criterion in judging interpersonal relations, so older ethical criteria – such as around monogamy, reproduction, marriage and non-instrumental incentives for sex (love) – lose legal regulative standing (as they often invoked ideas of duty over choice that would today be considered coercive). Have pornography and the contraceptive pill liberated people, or objectified the bodies of others (and ourselves) to be used for no higher purpose than personal (and immediate) physical gratification? Both are, in some sense, true. When is inciting arousal to be considered harmful? When is concealment good, and exposure obscene? It is hard to manage consent when acts are recorded and recordings then distributed. If revenge pornography highlights this problem most powerfully, the general question of how far participants are able to give consent for uses unknown has yet to be fathomed. Does graphic violence condone or encourage its emulation? Is simulated violence (such as in fiction films and television) less harmful because audiences know it is fake, and in terms of extreme pornography, does this distinction stand up? Do audiences think such violence is real, and, if so, is that better or worse than watching a boxing match, where the violence *is* real? Pornography has become mainstream (Presdee, 2000). One case of the search term 'rape' in 150,000 Pornhub titles does not suggest gender-based sexual violence has become mainstream, even as criminal violence towards male characters is within mainstream drama and video games. There are no simple answers to these questions. However, these questions highlight important complexities.

Online access makes certain forms of contact (grooming), content (child abuse imagery) and distribution crimes easier. However, as most child abuse occurs in domestic and local community settings, controlling children's ability to communicate with strangers may, in fact, reinforce the control of abusers; adult control over children's digital communication may be what enables abusers' evasion and concealment to be maintained through the silencing of victims. In terms of the relationship of digital accessibility to the production of harmful content, meanwhile, it can be argued that online access does not *directly* increase production as such, but, rather, that, because circulation may incite viewing of such material, greater access by viewers may, in practice, incite further production of abusive content.

Most sexually explicit images of 13–17 year olds are posted by 13–17 year olds. The internet has made it harder, but not impossible, to protect children from themselves. The idea of 'therapeutic jurisprudence', when applied to protecting children from themselves, is appropriate, but whether such an approach should be taken to those accessing such content is opposed by those keen to punish rather than cure. To protect children from adults that would do them harm is more readily agreed, and laws are more strict and better harmonized (though not fully) around the world today. Even as jurisdictional evasion and access are digitally afforded, so too is police action if resourced. Those accessing child pornography online tend to be social failures and isolates, who may benefit from therapy rather than punishment. Those who actually abuse children in real life tend to be more socially integrated. Teenagers sit between the space of familial protection and fully adult self-protection, and, as such, require guardians even as (or precisely because) they actively seek to avoid such paternalistic protections. Who, then, guards the guardians?

The age of consent at which a person can legally appear in pornography does vary across states, but significant harmonization has taken place over recent years precisely due to the challenge of online digital distribution in relation to national jurisdictions. The idea that distributed evasion (and concealment) makes regulating child pornography impossible is false. As Jewkes and Andrews (2005) point out, the paradox of online abuse imagery is that its circulation online actually makes concealment and evasion harder for perpetrators. This is for the same reason that such content that is in circulation is more harmful due to its reaching a wider audience (increasing the impact on victims and in extending the potential to normalize abuse for those viewing). While possession may not warrant the same criminal sanction in police 'hierarchies of standing' relative to production and distribution of child abuse imagery, possession alone within specific jurisdictions takes on increased significance in an age of global networks – even as collaboration between police forces is also facilitated by networks of communication and digital evidence trails. The same is true regarding online communications as digitally evidenced 'preparatory' crimes, such as in relation to grooming. The same tools that afford perpetrator access can also offer authorities the means of overcoming evasion and concealment.

Corruptions of citizenship

Citizens, states, corporations and hacktivists create and exploit the affordances of digital networks to maintain and infringe each other's privacy today. None are pure villains or guardians, as each actor plays both parts. The digital creates new domains for privacy to exist within, and new demands for such privacy – even as digital networks also afford intrusion and the arguments in favour of such intrusions. Demands for a 'contextual integrity' approach to privacy run up against new forms of context collapse. Increasingly subdivided lives demand new forms of information management (privacy), even as digital blurring of all such subsystem-specialized spaces (home, work, travel, shopping, intimacies and expression, for example) makes such separations and privacies harder to sustain. We want to overshare, and, at the same time, to protect our privacy ('the privacy paradox'). Debt, disclosure and desire circulate along with the forms of identity and identification that enable such circulation, making privacy harder to maintain; yet, rhizomic splitting sees recombining 'dividuals' exercise new forms of evasive and concealed access and incitement.

Trust in authorities (state and corporate) is reduced by revelations made by the very digital experts who they employ. State hackers come to moonlight as whistleblowers in today's increasingly outsourced and networked community of digital securocrats. At the same time, trust in ourselves is reduced when citizens and consumers find themselves split between a willingness to participate with Zuboff's (2015) 'Big Other's' data-mining and behavioural manipulating platforms, and the contrary desire not to be so surveilled and manipulated.

#MeToo and BLM's revelations breach the private/public divide, and networks like WikiLeaks and Anonymous do likewise in making whistleblowing easier. However, the image of the politically motivated hacker exposing the powerful has to be set against the ongoing work of tech-savvy geeks in creating the very surveillance infrastructure that some of these 'geeks' turned 'hackers' then expose. The genealogy of hacking, the ongoing contradictions between informating and automating, between state power, corporate profit and political progressiveness, and between hegemonically masculine hard mastery and libertarian egalitarianism, has no specific origin, logic or end.

The genealogy of unmediated media reveals ongoing contradictions. The lengthy concealment of the Zapruder film contrasts with increasingly rapid and widespread circulation of material today. From the camcorder footage of Rodney King's beating to phone footage of the 2004 Asian Tsunami, to later livestreamed beheadings and, later still, to the release of footage of George Floyd's murder, access has widened and accelerated. The question of whether unmediated media and citizen journalism represents increased truth or decontextualized information easily slotted into alternative narratives, relative to traditional (editorially controlled) media, remains disputed. Mass self-communication can reveal to wider publics or foster a narrowing of media consumption to within 'the daily me'. News online has become increasingly blended with entertainment and celebrity scandal, but this tendency towards 'human interest' and 'consumerism' in 'news' has been noted for over 200 years. Where digital access and concealed origin may afford criminal incitement, citizen journalism and whistleblowers can incite legitimate positive action.

Fake news is a problem, but not a new one. Today's 'disloyal fake news audience' (Nelson and Teneja, 2018) are not a coherent majority. Online news access remains dominated by traditional news sources, even when mediated by Facebook and Google. Time spent viewing traditional news content online is vastly greater than time spent viewing fake-news sources. The vast majority of those who access any fake-news sites spend far more time looking at established sources. Those that have a balanced online-news-media diet (the vast majority) are fairly good at discerning fake-news items. It is only a small minority that get drawn into near-exclusive fake-news bubbles. In particularly polarized political spaces (such as the United States), such bubbles compound hostilities where they exist. In more stable and integrated conditions, new media/citizen journalism and traditional forms of editorially controlled news media have found a productive balance between access and authority. Ironically, the overwhelming centrality of Facebook in terms of accessibility of fake news makes the regulation of such content online far easier than with a more plural traditional media, *if* such regulation is applied.

In other areas of media, such as in book publishing and academic journals, the tension between increased concentration of ownership and control, on the one hand, and the long tail of wider access for both audiences seeking information and writers seeking an audience, on the other, creates an unusual situation today. A small number of global cross-media corporations control an increasing proportion of media production (television, film, games, books, academic journals, magazines and so on); yet, a greater level of access to more content exists at the margins for those willing to look for it. Digital networks that afford free (IP infringing) access do more good than the supposed harm downloading/streaming are said to cause to existing business models.

Appropriations

Online access affords remote (evasive) and hidden (concealed, often micro-) fraud. Fraud, by its nature, involves incitement of the victim to give up money or goods by means of deception, and online concealment and evasion facilitate this deception. The rise of online transactions increases the need for identification to be alienable, that is, separable from the person identified, and in digital form. Such digital identifiers are designed to be transferable, creating scope for their fraudulent appropriation. Yet, forms of online encryption do work for most users most of the time, and institutional surveillance – in the form of 'chip and pin', security codes and, more recently, dual-factor authentication – limits the scope of fraud, even if it does not eliminate it. Technical securities, however, are limited as revealing security codes is essential in transacting with strangers, such that online banking and shopping create a point of transaction vulnerability that fraudsters can exploit.

Banking and shopping online do increase the risk of being the victim of online fraud, even as a number of precautions can reduce risk. The rise in online financial transactions predicts the rise in fraud online. Online transactions have vulnerabilities, just as any other form of interaction with strangers does. Before the internet, forms of identification also had to be alienable (for instance, signatures and photographs) and these could be faked. Online fraud operates faster and further, however, but then so does digital verification. Counterfeit credit cards (where no real account matched the card) are largely useless today as remote digital verification is required and is nearly instantaneous. Cloned card fraud has arisen to replace counterfeit cards, but this is itself limited in use by the dual-factor checks that near-universal use of mobile, digital, network devices now affords.

Low self-control is a weak predictor of becoming a victim; scale of online transaction is the strongest. Self-confidence that one can manage risk increases risky behaviour. Blaming victims is doubly counter-productive as it both blinds the overly self-confident and reduces the willingness of victims to report, masking the problem. Cross-border police actions have successfully targeted some forms of romance (and other fraudulent) scams, even as most fraud requires a local point of payoff that reduces scope for geographical evasion and concealment. The key issue is not technological, but, rather, the resourcing of anti-fraud action, as well as problem ownership among numerous state, private and personal guardians.

Downloading or streaming copyright-protected content (music, film, computer software and live sports coverage) transforms infringing access relative to any previous mode of copying and distributing. As broadband speeds increased, so the speed of downloading and/or streaming such material increased. Content that once required dedicated formats (records, tapes and CDs, or specialized satellite/cable channels) became readily accessible via faster personal internet connections. The rise of distributed peer-to-peer software, then torrent-based sharing platforms, and then streaming channels made it increasingly possible for various parties to copyright infringing to conceal themselves from legal targeting. VPNs and remote servers made it possible to evade legal strategies and technical forms of blocking. While directly inciting viewers to access copyrighted content remains illegal, dual-use technologies that enable users to infringe copyright are legal to distribute if legal uses exist (such as recording public service materials or downloading content that is out of copyright).

Free access to previously expensive, copyrighted content has incited major changes in user behaviour, but also in content-industry business models. Music fans pay more for live concert tickets, and the volume of tickets sold rocketed when opportunity costs (from buying recordings) fell. Record companies have been forced to licence alternative ways of distributing music in the face of more efficient (and free) online alternatives to physical copies (first, pay-per-song services like iTunes, and then advertiser/subscription services like Spotify). Cinema attendances/home cinema subscriptions have risen even as video rentals collapsed in the face of free online downloading and streaming. The encoded nature of live digital sports television saw the generation of highly profitable subscription services. However, once high-quality, free, streaming services became available, it was harder to grow demand for subscription services (cementing the dominant position of first movers in the commercially encrypted marketplace). Even in the face of (access to) the incitement offered by illegally free alternatives (alternatives that can evade authorities and conceal infringement), free and legal alternative business models have arisen. The case of computer games and software, fields where no live alternative exists, illustrates just how far technical developments can maintain audience willingness to pay, even when free alternatives exist for those that do not have to, cannot or will not pay.

Access

Access by means of the 'long fuse' or 'logic bomb' has not come to pass as a significant terrorist threat. The spectacle of the suicide attack has risen to prominence precisely with the rise of global digital media networks. While network access does create a sense that there is nowhere to hide from digital bullies, stalkers and trolls, the digital also creates scope to find one's tribe and, therefore, new kinds of safe spaces online. Online access to pornography is pervasive, and increasingly legal (in number of jurisdictions and scope of content); yet, legislation and forms of blocking have developed so there is not a free-for-all. Online grooming and wider circulation of child abuse imagery are significant problems, but surveillance also increases by authorities, even as victims in real-world contexts have greater ability to speak out when traditional guardians fail them. Invasion of privacy, by state, corporate and criminal actors, is afforded by digital network access. Nonetheless, securocratic geeks turned hackers, moonlighting as whistleblowers, reverse the gaze, highlighting the ongoing instability of the hacker genealogy. Demands for privacy, while often paradoxical, have not gone away in an age of digital surveillance, and, in many ways, they intensify precisely in the face of potential (and awareness of the potential for) intrusion. The unstable genealogy of unmediated media and citizen journalism, and the struggle between authority and expression, censorship and fake news, continues to unfold. Online banking and shopping require alienable digital identifiers and so create new scope for fraud, but the very channels that enable remote access by fraudsters also enable faster tracking and authentication of transactions by banks and credit card companies. Network access to free downloads of digital content have also led to the creation of easy-to-access, legal, services – many of which are also free to use.

Concealment

The suggestion that terrorist networks are able to communicate secretly online via encrypted channels in a fashion far more private than any previous method has proven misguided. States have proven far better at surveilling terrorists than terrorists have proven capable of planning in secret via online channels. While bullies, stalkers and trolls may create the impression of invisibility in their digital modes of intrusion and harassment, digital mouse droppings and the 'disappearance of disappearance' create data trails in excess of those produced in real-world manifestations of such abusive behaviour. The viewer of pornography online gains a degree of privacy relative to real-world modes of accessing such content, but, as noted, such online access creates a greater data trail than might shopping for such content with cash in a shop. This creation of a trail is all the more true in relation to child pornography, where police hierarchies of standing have made tracking those accessing such content online a greater priority in recent years, and where possession has taken on greater legal significance where local law enforcement seek to target those within their immediate jurisdiction. The cat-and-mouse struggle between encryption and surveillance in relation to personal privacy is not resolved one way or the other in absolute terms, but rather depends on the resources deployed by competing actors in tracing or evading the circulations of debt, disclosure and desire in online spaces. Hidden algorithms seek to direct and persuade us in our choice of viewing, relative to more explicit editorial controls in past times, and in relation to traditional forms of media production; yet, today's disloyal, online fake-news audience is not as narrow minded as has often been suggested – at least in most cases, most of the time, in most places. Nonetheless, in polarized social settings, manipulation has more significant impacts. Online micro-frauds, meanwhile, rely on the capacity of virtual bank robbery to be less visible than its real-world equivalents; but online surveillance and digital verification make such intrusions harder to conceal than in the days of counterfeit credit cards. Peer-to-peer sharing networks, torrent services and streaming make it harder to identify IPRs infringing copiers, but the targeting of uploaders, downloaders or service providers still seeks to regulate such practices.

Evasion

The suggestion that terrorist groups today better adapt to digital networks, as viral agents, relative to hierarchical and bureaucratic states and their military structures, has not proven to be correct. Digilantism, as a form of engagement with online trolls, bullies and stalkers, has shown some success in breaking criminal evasion – even as remote intrusion can make evasion easier, at least until guardians apply sufficient resources to reveal abusers. Cooperation between law enforcement agencies limits scope to evade the law in relation to pornography, even if, in some part, harmonization of laws between states reflects a relaxation of obscenity laws in many jurisdictions in the face of pervasive online access – and the ability to evade legal and technical forms of prevention. Regarding child pornography, harmonization has gone further, as has a greater consensus over harm and the threat posed by the circulation of child abuse imagery. VPNs and TORs reduce scope for state surveillance and control over communications, as do encrypted services like WhatsApp and Telegram. However, levels of corporate intrusion into privacy, via services like Facebook, TikTok and Zoom, and by state-based digital surveillance agencies, remain massive and hard to avoid, unless one takes very serious care, which most people simply do not do. Action at a distance does increase the scope for online fraudsters to evade the law, but payment usually requires some form of local conduit creating vulnerability. Blocking access to copyright-infringing services can be worked around, but this does increase the effort required to evade controls, and with free legal streaming alternatives, the cost-benefit of infringement can be managed.

Incitement

Beheading videos did not have the widespread effects commonly attributed to them, but did have localized impact. Online propaganda has reinforced the radicalization of some, but state and mainstream messaging remains dominant. The low-tech lone wolf reflects the failure of online radicalization as a force multiplier far more than it evidences the success of such efforts at terrorist mobilization. Social media has democratized a certain form of following, intrusion and uninvited communication, but has not incited a radical transformation in the nature of the most harmful forms of such action. However, online tribes do foster a furthering of polarization between communities of like-minded dislike for those who do not agree with them.

Meanwhile, it is not the case that pornography online has created a more misogynistic and abusive society (and, indeed, may afford the exact opposite). However, extreme pornography can normalize and reinforce existing hostilities where they exist. Child pornography online may have similar harmful disinhibiting effects on a particular type of person, even as the spread of such content has also triggered significant campaigning and legislative change that has strengthened and harmonized laws in most countries in recent years. Digital intrusion into personal privacy by states and corporations has led to increased calls for tighter privacy protections, even while most people practice a paradoxical mix of concern in principle and indifference in practice (whether because people feel powerless to act or because they choose not to exercise what power they do have). Whistleblowers previously employed to perform that which they then bring to light, again, manifest a tension between symmetry and asymmetry: asymmetry between what digital experts know, when most of us rely on them to tell us; and yet symmetry in terms of who discloses what to whom (such that those employed by states and corporations to monitor citizens and customers) also uses digital affordances to reveal such practices. Such hackers are both gamekeeper and poacher – even if exactly which role is which becomes confusing when law enforcers become the primary agents of infringement, and those highlighting such infringement are criminalized. The ecosystems of traditional and new-media news generation and distribution are complex in relation to who is persuaded and what they are persuaded of. Bubbles emerge and can be reinforced by fake news circulated by unregulated channels; but other bubbles have been burst by means of challenger/citizen reporting, revealing that which might otherwise have gone unreported. Fear of loss, or false hope (of romance), can be incited by extortionists/fraudsters online, but guardianship in online spaces has not disappeared and can also be extended by such networks. Free content has incited infringement, but new network channels of distribution have also incited new business models that have reinvented the creative industries.

Balancing risks

The affordances of digital networks are symmetrical in principle but asymmetrical in practice, such that it is a question of *resources* and *priority* that determines outcomes, not technological logic (whatever that might be). Resources can be deployed, and, when they are, significant degrees of protection can be afforded by the very networks that can afford harm; but the question of priority is perhaps the more complex one. As this book has highlighted, there are harms that can arise from too much surveillance and policing online, just as there are harms that can arise from there being too little regulation. The question of who guards the guardians must always be asked, even as we continue to ask whether the internet reduces guardianship and hence affords an increase in criminal harm by means of access, evasion, concealment and/or incitement. Panic over cyber terror, hate speech, online trolls and stalkers, revenge and child pornography, fake news and stalker intrusion, fraud and downloading can create an atmosphere ripe for demanding that authorities get tough, and, in such conditions, free expression, user privacy, innovation and creativity may be quashed in the name of ever greater levels of 'protection'.

Regarding politically motivated and personal 'hate' speech, the potential for the internet to circulate abusive communication and, thereby, to incite harm, whether in the form of radicalization, intimidation or humiliation, is real. At the same time, the capacity to be offended by those who disagree with an individual or a group should not be sufficient grounds for claiming such disagreement constitutes incitement to criminal hatred. John Stuart Mill's (2008 [1859]) defence of free expression was limited by his belief that such freedom should be curtailed where expression incites harm. However, Mill did not clearly set out where harm ends and simply being offended begins. As a society, we are therefore compelled to draw that line for ourselves, and we will have to accept that we will never fully agree on how to do so.

Regarding obscenity, again, the line between freedom of expression and exploration, on the one hand, and, on the other, content that might cause harm (whether to participants, viewers or those that viewers might themselves then harm) is not a simple one. Academic dispute over 'media effects' has raged for over a century, with no sign of being resolved. Parties to this debate tend to reinforce their own views rather than arrive at any sort of consensus. While it is possible to regulate online content and harmonize laws to a significant degree, what remains impossible to achieve is a consensus on exactly what should and should not be allowed. Deciding that 'extreme', 'revenge' and 'child' pornography are bad is the easy part; deciding what necessarily counts as each is not so easy. Some consensus has arisen, and significant harmonization has taken place, but the tension between censorship and expression will remain, and should do so, as neither side is absolutely 'right'. That some content considered obscene by some people remains online might seem like a bad thing, but if nothing anyone considered obscene existed online, perhaps that would be worse. Dispute is better than silence. Efforts to resolve disputes should still be sought. That will mean compromise.

The invasion of privacy by states and corporations has been revealed by whistleblowers breaching the privacy of those who once paid them to do precisely that to others. This is perhaps the best of all possible worlds. A world without privacy and one where privacy was absolute would be equally problematic. Symmetrical panics about too much or too little privacy seem always to compel 'tougher' action – one way and then the other – so as to take us right back to where we already were. Unmediated media can challenge censorship to reveal the truth, or circulate fake news that undermines the very idea of truth. Nevertheless, in conducive social conditions, the balance between new and old media has created a news ecosystem that tempers the limits of both editorial/algorithmic and unmediated biases. We may simply need to learn to live with some anticipation of error, even as that also means learning to spot it, not just accept it. Rejecting blind faith as well as maintaining an awareness of uncertainty, alongside the need to evaluate critically what we are told, does not mean a collapse into cynicism.

Protection against fraud requires a level of surveillance over our transactions by state and financial institutions that we may be willing to accept, even as the freedom to transact with strangers may involve a level of risk that people must also be willing to accept. In like fashion, the dual-use capabilities of technologies that afford copyright infringement should be accepted if we want to benefit from innovation. Adaptation to such challenging conditions compels creative industries to adapt. Leaky systems may, in the end, be more dynamic than systems designed first and foremost to nail down and control every aspect of property and exchange.

At the end of the day, the balance between freedom and risk is a question of social and political priorities, not technical affordances and jurisdictional reach. We need to make choices about the choices we make and which we should be allowed to make. The desire to eliminate risk needs to be tempered by the desire to remain free enough to take such risks. Freedom will create spaces in which some victims will experience some harm. This is an imperfect world. The creation of a world where everyone is protected from everything all of the time would be worse – as there would be nowhere to hide from those who would lock us up in the name of protection from harm.

References

Abbate, J. (1999) *Inventing the Internet*, Cambridge, MA: MIT Press.

Acquardro, M. and Begotti, T. (2019) 'Prevalence of Cyberstalking and Previous Offline Victimization in a Sample of Italian University Students', *Social Sciences*, 8(1): 1–10.

Acquisti, A., Brandimante, L. and Loewenstein, G. (2015) 'Privacy and Human Behaviour in the Age of Information', *Science*, 347(6221): 509–514.

Adams, W. and Flynn, A. (2017) Federal Prosecution of Commercial Sexual Exploitation of Children Cases 2014–2013, US Department of Justice: Special Report, October, [online]. Available from: https://nmc sap.org/wp-content/uploads/Prosecution_of_Commercial_Sexual_Exploitation_2004-2013.pdf [Accessed 18 July 2022].

Agger, H. (2015) *Oversharing: Presentations of Self in the Internet Age*, London: Routledge.

Aguiar, L. and Martens, B. (2016) 'Digital Music Consumption on the Internet: Evidence from Clickstream Data', *Information Economics and Policy*, 34: 27–43.

Ahlgrim, B. and Terrance, C. (2018) 'Perceptions of Cyberstalking', *Journal of Interpersonal Violence*, 36(7–8): 4074–4093.

Akdeniz, Y. (2010) 'To Block or Not to Block: European Approaches to Content Regulation and Implications for Freedom of Expression', *Computer Law and Security Review*, 26(3): 260–272.

Albahar, M. (2017) 'Cyber attacks and Terrorism: A Twenty First Century Conundrum', *Science and Engineering Ethics*, 25: 993–1006.

Albini, S. (1993) 'The Problem with Music', *The Baffler*, [online] 5. Available from: https://thebaffler.com/salvos/the-problem-with-music [Accessed 9 December 2021].

Alexander, I. (2007) 'Criminalizing Copyright: A Story of Publishers, Pirates and Pieces of Eight', *Cambridge Law Journal*, 66(3): 625–656.

Allan, S. (2007) 'Citizen Journalism and the Rise of "Mass Self-Communication"', *Global Media Journal*, 1(1): 1–20.

Allen, S. (2013) *Citizen Witnessing: Revisioning Journalism in Times of Crisis*, Cambridge: Polity Press.

Allen, S. and Thorsen, E. (2009) *Citizen Journalism*, Bern: Peter Lang.

Amoore, L. (2014) 'Security and the Claim to Privacy', *International Political Sociology*, 8(1): 108–112.

Anderson, C. (2009) *The (Longer) Long Tail* (undated and expanded edition), London: Random House.

Anderson, C., Bell, E. and Shirky, C. (2015) 'Post-Industrial Journalism: Adapting to the Present', *Geopolitics, History, and International Relations*, 7(2): 32–123.

Andregg, M. (2016) 'Ethical Implications of the Snowden Revelations', *International Journal of Intelligence, Security and Public Affairs*, 18(2): 110–131.

Andrejevic, M. (2002) 'The Kinder, Gentler Gaze of Big Brother: Reality TV in the Era of Digital Capitalism', *New Media and Society*, 4(2): 251–270.

Andrejevic, M. and Gates, K. (2014) 'Big Data Surveillance: Introduction', *Surveillance and Society*, 12(2): 185–196.

Arthur, C. (2013) LulzSec: What they did, who they were and how they got caught. *The Guardian*, [online], 16 May. Available from: www.theguardian.com/technology/2013/may/16/lulzsec-hacking-fbi-jail [Accessed 28 March 2022].

Arvanitakis, J. and Fredriksson, M. (2016) 'Commons, Piracy and the Crisis of Property', *TripleC: Communication, Capitalism, Critique*, 14(1): 132–144.

Atkinson, R. and Rodgers, T. (2016) 'Pleasure Zones and Murder Boxes: Online Pornography and Violent Video Games as Cultural Zones of Exception', *British Journal of Criminology*, 56: 1297–1307.

Attwood, F. and Smith, C. (2010) 'Extreme Concern: Regulating "Dangerous Pictures" in the United Kingdom', *Journal of Law and Society*, 37(1): 171–188.

Attwood, F. and Walters, C. (2013) 'Fifty Shades and the Law: Regulating Sex and Sex Media in the UK', *Sexualities*, 16(8): 974–979.

Atwan, A.B. (2015) *Islamic State: The Digital Caliphate*, Oakland: University of California Press.

Awan, I. (ed) (2016) *Islamophobia in Cyberspace: Hate Crimes Go Viral*, London: Routledge.

Awan, I. (2017) 'Cyber-extremism: Isis and the Power of Social Media', *Society*, 54(2): 138–149.

Baghai, K. (2012) 'Privacy as a Human Right: A Sociological Theory', *Sociology*, 46(5): 951–965.

Bakir, V. and McStay, A. (2018) 'Fake News and the Economy of Emotions', *Digital Journalism*, 6(2): 154–175.

Bakkar, P. and Taalas, S. (2007) 'The Irresistible Rise of Porn: The Untold Story of a Global Industry', *Observatorio*, 1(1).

Balmas, M. (2014) 'When Fake News Becomes Real', *Communication Research*, 41(3): 430–454.

Bambauer, D.E. (2013) 'Privacy versus Security', *The Journal of Criminal Law and Criminology*, 103(3): 667–683.

Bandura, A. (1963) 'Imitation of Film-mediated Aggression Models', *The Journal of Abnormal and Social Psychology*, 66(1): 3–11.

Banks, J. (2010) 'Regulating Hate Speech Online', *International Review of Law, Computers and Technology*, 24(3): 233–239.

Barth, S. and de Jong, M. (2017) 'The privacy paradox', *Telematics and Informatics*, 34(7): 1038–1058.

Bauman, Z., Bigo, D., Esteves, P., Guild, E., Jabri, V., Lyon, D. and Walker, R. (2014) 'After Snowden: Rethinking the Impact of Surveillance', *International Political Sociology*, 8(2): 121–144.

Beck, U. (1992) *Risk Society: Towards a New Modernity*, London: Sage.

Beckett, C. and Ball, J. (2012) *WikiLeaks: News in the Networked Era*, Cambridge: Polity Press.

Benkler, Y. (2011) 'A Free Irresponsible Press: WikiLeaks and the Battle over the Soul of the Networked Fourth Estate', *Harvard Civil Rights – Civil Liberties Law Review*, 46(2): 311–397.

Benson, D.C. (2014) 'Why the Internet is not Increasing Terrorism', *Security Studies*, 23(3): 293–328.

Benthall, S., Seda, G. and Nissenbaum, H. (2017) 'Contextual Integrity Through the Lens of Computer Science', *Foundations and Trends in Privacy and Security*, 2(1): 1–69.

Bentley, H., Fellowes, A., Glenister, S., Mussen, N., Ruschen, H., Slater, B., Turnbull, M., Vine, T., Wilson, P. and Witcombe-Hayes, S. (2020) *How Safe Are Our Children?: An Overview of Data on Adolescent Abuse*. London: NSPCC.

Berghel, H. (2017) 'Lies, Damn Lies, and Fake News', *Computer*, 50(2): 80–85.

Bernat, B. and Makin, D. (2014) 'Cybercrime Theory and Discerning if There Is a Crime: The Case of Digital Piracy', *International Review of Modern Sociology*, 40: 99–119.

Berners-Lee, T. (2000) *Weaving the Web: The Original Design and Ultimate Destiny of the World Wide Web*, San Francisco, CA: Harper Collins.

Berry, M.J. and Bainbridge, S.L. (2017) 'Manchester's Cyberstalked 18–30s', *Advances in Social Science Research Journal*, 4(18): 73–85.

Birmingham, J. and David, M. (2011) 'Live-streaming: Will Football Fans Continue to Be More Law Abiding Than Music Fans?', *Sport in Society*, 14(1): 69–80.

Blanke, J.M. (2018) 'Privacy and Outrage', *Journal of Law, Technology and the Internet*, 9(1): 1–17.

Bleich, E. (2011) 'The Rise of Hate Speech and Hate Crime Laws in Liberal Democracies', *Journal of Ethnic and Migration Studies*, 37(6): 917–934.

Bocij, P. and MacFarlane, L. (2003) 'Cyberstalking: The Technology of Hate', *The Police Journal*, 76(3): 204–221.

Bonanno, R.A. and Hymel, S. (2013) 'Cyber Bullying and Internalizing Difficulties', *Journal of Youth and Adolescence*, 42(5): 685–697.

Borja, K., Dieringer, S. and Daw, J. (2015) 'The Effect of Music Streaming Services on Music Piracy Among College Students', *Computers in Human Behaviour*, 45: 69–76.

Bossler, A.M. and Holt, T.J. (2009a) 'The Effects of Self-Control on Victimization in the Cyberworld', *Journal of Criminal Justice*, 38(1): 227–236.

Bossler, A.M. and Holt, T.J. (2009b) 'On-line Activities, Guardianship and Online Infections: An Examination of Routine Activities Theory', *International Journal of Cyber Criminology*, 3(1): 400–420.

Bozarova, N. (2012) 'Public Intimacy: Disclosure Interpretation and Social Judgement on Facebook', *Journal of Communication*, 62(5): 815–832.

BPI (2008) *More than the Music: The UK Recorded Music Business and Our Society,* London: BPI.

Brenner, S. (2007) '"At Light Speed": Attribution and Responses to Cybercrime/Terrorism/Warfare', *The Journal of Criminal Law and Criminology*, 97(2): 379–475.

Brevini, B. (2017) 'WikiLeaks: Between Disclosure and Whistle-blowing in Digital Times', *Sociology Compass*, 11(3): 1–11.

Brochado, S., Soares, S. and Fraga, S. (2016) 'A Scoping Review on Studies of Cyber-bullying Prevalence Among Adolescents', *Trauma, Violence and Abuse*, 18(5): 523–531.

Brown, I. (2015) 'Copyright Technologies and Clashing Rights', in M. David and D. Halbert (eds) *The Sage Handbook of Intellectual Property*, Sage: London, pp 567–585.

Brown, I. and Kraft, D. (2009) 'Terrorism and the Proportionality of Internet Surveillance', *European Journal of Criminology*, 6(2): 119–134.

Brown, S. (2014) 'Porn Piracy: An Overlooked Phenomenon', *Porn Studies*, 1(3): 326–330.

Bryce, J. (2009) 'Online Sexual Exploitation of Children and Young People', in Y. Jewkes and M. Yar (eds) *Handbook of Internet Crime*, Cullumpton: Willan, pp 320–342.

Buckley, B. and Hunter, M. (2011) 'Say Cheese! Privacy and Facial Recognition', *Comparative Law and Security Review*, 27(6): 637–46.

Business Insider (2021) US Digital Banking Users Will Surpass 200 Million in 2022 [online]. Available from: www.businessinsider.com/current-state-of-online-banking-industry?r=US&IR=T [Accessed 30 November 2021].

Butter, M. (2020) *The Nature of Conspiracy Theories*, Cambridge: Polity Press.

Calvert, C. (2015) 'Revenge Porn and Freedom of Expression: Legislative Pushback to an Online Weapon of Emotional and Reputational Destruction', *Fordham Intellectual Property Media and Entertainment Law Journal*, 24(3): 673–702.

Cantijoch, M. (2013) 'Read the Riots: What Were the Police Doing on Twitter', *Policing and Society*, 23(4): 413–436.

Carpenter, D., Nah, S. and Chung, D. (2015) 'A Study of Community Journalists and Their Organisational Characteristics and Story Generating Routine', *Journalism* 16(4): 505–520.

Carpenter, S. (2010) 'A Study of Content Diversity in Online Citizen Journalism and Online Newspaper Articles', *New Media and Society*, 12(7): 1066–1084.

Carr, J. (2004) *Child Abuse, Child Pornography and the Internet*, London: NCH, The Children's Charity.

Carson, R. (1962) *Silent Spring*, Boston, MA: Houghton Mifflin.

Carter, A. (1985) *Come Upon these Yellow Sands*, London: Bloodaxe Books.

Carter, D. and Rogers, I. (2014) 'Fifteen Years of "Utoptia": Napster and Pitchfork as Technologies of Democratisation', *First Monday*, 19(10).

Casadesus-Masanell, R. and Hervas-Drane, A. (2010) 'Competing Against Online Sharing', *Management Decision*, 48(8): 1247–1260.

Castells, M. (1996) *The Rise of the Network Society*, Oxford: Blackwell.

Castells, M. (2009) *Communication Power*, Oxford: Oxford University Press.

Castells, M. (2012) *Networks of Outrage and Hope*, Cambridge: Polity.

Castells, M., Fernandez-Arderol, M., Linchman Qiu, J. and Sey, A. (2007) *Global Communication and Society*, Cambridge, MA: Massachusetts Institute of Technology Press.

Cavelty, M.D. (2008) 'Cyber-Terror – Looming Threat or Phantom Menace? The Framing of the US Cyber-Threat Debate', *Journal of Information Technology and Politics*, 4(1): 19–36.

Choi, K., Lee, C.S. and Cadigan, R. (2018) 'Spreading Propaganda in Cyberspace: Comparing Cyber-Resource Usage of Al Qaeda and ISIS', *International Journal of Cybersecurity Intelligence & Cybercrime*, 1(1): 21–39.

Christiansen, C. (2014) 'WikiLeaks and the Afterlife of Collateral Murder', *International Journal of Communication*, 8: 2593–2602.

Citron, D.K. and Norton, H. (2011) 'Intermediaries and Hate Speech', *Boston University Law Review*, 91: 1435–1484.

Clough, J. (2015) *Principles of Cybercrime*. Cambridge: Cambridge University Press.

Cohen, S. (1973) *Folk Devils and Moral Panics: The Creation of the Mods and Rockers*, London: Paladin.

Cohen, S. (1985) *Visions of Social Control: Crime, Punishment and Classification*, Chichester: Wiley.

Cohen, L. and Felson, M. (1979) 'Social Change and Crime Rate Trends: A Routine Activity Approach', *American Sociological Review*, 44(4): 588–608.

Cohen-Almagor, R. (2014) 'Countering Hate on the Internet', *Annual Review of Law and Ethics*, 22: 1–21.

Coleman, G. (2011) 'Hacker Politics and Publics', *Public Culture*, 23(3): 511–516.

Coleman, G. (2014) *Hacker, Hoaxer, Whistleblower, Spy: The Many Faces of Anonymous*, London: Verso.

Coleman, G. (2019) 'How Has the Fight for Anonymity and Privacy Advanced Since Snowden's Whistleblowing?', *Media, Culture and Society*, 41(4): 565–571.

Connell, R. and Messerschmidt, J. (2005) 'Hegemonic Masculinity: Rethinking the Concept', *Gender and Society*, 19(6): 829–859.

Conte, J. (1984) 'The Justice System and Sexual Abuse of Children', *Social Service Review*, 58(4): 556–568.

Conway, M. (2011) 'Against Cyberterrorism', *Communications of the ACM*, 54(2): 26–28.

Craigie-Williams, G. (2018) 'The Rise of Citizen Journalism and Fake News', *Medium*, [online] 7 January. Available from: https://medium.com/@journalismblog/the-rise-of-citizen-journalism-and-fake-news-a643b318591c [Accessed 28 March 2022].

Craven, S., Brown, S. and Gilchrist, E. (2007) 'Current Responses to Sexual Grooming: Implications for Prevention', *The Howard Journal of Crime and Justice*, 46(1): 60–71.

Cross, C. (2013) 'Nobody's Holding a Gun to Your Head: Examining Current Discourses Surrounding Victims of Online Fraud', in K. Richards and J. Tauri (eds) *Crime, Justice and Social Democracy*, Brisbane: Queensland University of Technology, pp 25–32.

Cross, C. (2015) 'No Laughing Matter: Blaming the Victim of Online Fraud', *International Review of Victimology*, 21(2): 187–204.

Cross, C. and Blackshaw, D. (2015) 'Improving the Police Response to Online Fraud', *Policing*, 9(2): 119–128.

Cunningham, S., Engelstätter, B. and Ward, M.R. (2016) 'Violent Video Games and Violent Crime', *Southern Economic Journal*, 82(4): 1247–1265.

Curien, N. and Moreau, F. (2009) 'The Music Industry in the Digital Era: Towards New Contracts', *Journal of Media Economics*, 22(2): 102–13.

Currah, A. (2007) 'Managing Creativity: The Tensions Between Commodities and Gifts in a Digital Networked Environment', *Economy and Society*, 36(3): 467–494.

Da Rimini, F. (2013) 'The Tangled Hydra: Developments in Transglobal Peer-to-peer Culture', *Global Networks*, 13(3): 310–329.

David, M. (2005) *Science in Society*, London: Palgrave.

David, M. (2006) 'Romanticism, Creativity and Copyright: Visions and Nightmares', *European Journal of Social Theory*, 9(3): 425–433.

David, M. (2010) *Peer to Peer and the Music Industry*, London: Theory, Culture and Society Monograph Series.

David, M. (2011) 'Music Lessons: Football Finance and live streaming', *Journal of Policy Research in Tourism, Leisure and Events*, 3(1): 95–98.

David M. (2013) 'Cultural, Legal, Technical, and Economic Perspectives on Copyright Online: The Case of the Music Industry', in W. Dutton (ed) *The Oxford Handbook of Internet Studies*, Oxford: Oxford University Press, pp 464–485.

David, M. (2016) 'The Legacy of Napster', in R. Nowak and A. Whelan (eds) *Networked Music Cultures: Contemporary Approaches, Emerging Issues*, Basingstoke: Palgrave Macmillan, pp 49–65.

David, M. (2017a) 'Sharing: Post-Scarcity Beyond Capitalism?', *Cambridge Journal of Regions, Economy and Society*, 10(2): 311–325.

David, M. (2017b) *Sharing: Crime Against Capitalism*, Cambridge: Polity Press.

David, M. (2019a) 'CD', in C. Op Den Kamp and D. Hunter (eds) *A History of Intellectual Property in 50 Objects*, Cambridge: Cambridge University Press, pp 361–367.

David, M. (2019b) 'Incentives to Share in the Digital Economy', in M. Graham and W. Dutton (eds) *Society and the Internet: How Networks of Information and Communication are Changing Our Lives*, Oxford: Oxford University Press, pp 323–337.

David, M. and Halbert, D. (2015) *Owning the World of Ideas*, London: Sage.

David, M. and Halbert, D. (2017) 'Intellectual Property & Global Policy', *Global Policy*, 8(2): 149–158.

David, M. and Kirkhope, J. (2006) 'The Impossibility of Technical Security: Intellectual Property and the Paradox of Informational Capitalism', in M. Lacy and P. Witkin (eds) *Global Politics in an Information Age*, Manchester, Manchester University Press, pp 80–95.

David, M. and Millward, P. (2012) 'Football's Coming Home?: Digital Reterritorialization, Contradictions in the Transnational Coverage of Sport and the Sociology of Alternative Football Broadcasts', *British Journal of Sociology*, 63(2): 349–369.

David, M., Kirton, A. and Millward, P. (2015) 'Sports Television Broadcasting and the Challenge of Live-Streaming', in M. David and D. Halbert (eds) *The Sage Handbook of Intellectual Property*, London: Sage, pp 435–450.

David, M., Kirton, A. and Millward, P. (2017) 'Castells, "Murdochization", Economic Counterpower and Livestreaming', *Convergence: The International Journal of Research into New Media Technologies*, 23(5): 497–511.

Davies, N. (2008) 'Churnalism Has Taken the Place of What We Should Be Doing: Telling the Truth', *The Press Gazette*, [online] 4 February. Available from: www.pressgazette.co.uk/nick-davies-churnalism-has-taken-the-place-of-what-we-should-be-doing-telling-the-truth-40117/ [Accessed 22 November 2021].

Davis, M. (2007) *Buda's Wagon: A Brief History of the Car Bomb*, New York, Verso.

Davis, K. and James, C. (2013) 'Tweens' Conceptions of Privacy Online', *Learning, Media and Technology*, 38(1): 4–25.

DeCamp, W. and Ferguson, C.F. (2017) 'The Impact of Degree of Exposure to Violent Video Games, Family Background, and Other Factors on Youth Violence', *Journal of Youth and Adolescence*, 46(2): 388–400.

Decary-Hetu, D. and Dupont, B. (2012) 'The Social Network of Hackers', *Global Crime*, 13(3): 160–175.

Decary-Hetu, D., Morselli, C. and Leman-Langlois, S. (2012) 'Welcome to the Scene: A Study of Social Organisation and Recognition among Warez Hackers', *Journal of Research in Crime and Delinquency*, 49(3): 359–382.

Deleuze, G. (1992) 'Postscript on the Societies of Control', *October*, 59(Winter): 3–7.

Deleuze, G. and Guattari, F. (1984) *Anti-Oedipus: Capitalism and Schizophrenia*, London: Athlone.

Delmas, C. (2015) 'Whistleblowing', *Social Theory and Practice*, 41(1): 77–105.

Delmas, C. (2017) 'Is Hacktivism the New Civil Disobedience?', *Raisons Politiques*, 69(1): 63–81.

Denham, J. and Spokes, M. (2019) 'Thinking Outside the "Murder Box": Virtual Violence and Pro-social Action in Video Games', *The British Journal of Criminology*, 59(3): 737–755.

Denning, D. (2000) '"Cyberterrorism", Testimony before the Special Oversight Panel of Terrorism Committee on Armed Services', US House of Representatives, [online] 23 May. Available from: https://faculty.nps.edu/dedennin/publications/Testimony-Cyberterrorism2000.htm [Accessed 14 December 2022].

Denning, D. (2007) 'A View of Cyberterrorism 5 Years Later', in K.E. Himma (ed) *Internet Security: Hacking, Counterhacking and Society*, London: Jones and Bartlett, pp 123–139.

Depoorter, B. (2014) 'What Happened to Video Game Piracy?', *Communications of the ACM*, 57(5): 33–34.

Derwenter, R., Haucup, J. and Wenzel, T. (2012) 'On File Sharing with Indirect Network Effects Between Concert Ticket Sales and Music Recording', *Journal of Media Economics*, 25(2): 168–178.

Deseriis, M. (2016) 'Hacktivism: On the Use of Botnets in Cyber Attacks', *Theory, Culture and Society*, 34(4): 131–152.

Determann, L. and Guttenberg, K. (2014) 'On War and Peace in Cyberspace: Security, Privacy, Jurisdiction', *Hastings Constitutional Law Quarterly*, 41(4): 875–902.

Dizon, M. (2019) *A Socio-Legal Study of Hacking: Breaking and Remaking Law and Technology*, New York: Routledge.

Dorf, M. and Tarrow, S. (2017) 'Stings and Scams: "Fake News"', *Journal of Constitutional Law*, 20(1): 1–32.

Douglas, M. (1992) *Risk and Blame: Essays in Cultural Theory*, London: Routledge.

Durkheim, E. (1975) *Emile Durkheim on Morality and Society*, Chicago: University of Chicago Press.

DW Documentary (2020) 'WikiLeaks – Public Enemy: Julian Assange', *Made for Minds*, [online] 25 February. Available from: www.dw.com/en/wikileaks-public-enemy-julian-assange/av-52526807 [Accessed 28 March 2022].

Economist, The (2012) 'Online software piracy: Head in the Clouds', *The Economist*, [online] 25 July. Available from: www.economist.com/blogs/graphicdetail/2012/07/online-software-piracy [Accessed 27 March 2022].

Elias, N. (1994) *The Civilizing Process*, Oxford: Blackwell.

Embar-Seddon, A. (2003) 'Cyberterrorism: Are We Under Siege?', *American Behavioral Scientist*, 46(6): 1033–1043.

Fenton, N. (2010) 'Drowning or Waving? New Media, Journalism and Democracy', in N. Fenton (ed) *New Media, Old News*, London: Sage, pp 3–16.

Ferguson, C.F. (2010) 'Blazing Angels or Resident Evil? Can Violent Video Games be a Force for Good?', *Review of General Psychology*, 14(2): 68–81.

Ferrall, J., Hayward, K. and Young, J. (2015) *Cultural Criminology: An Invitation*, London: Sage.

Finnegan, H. and Fritz, P. (2012) 'Differential Effects of Gender on Perception of Stalking and Harassment Behaviour', *Violence and Victims*, 27(6): 895–910.

Fortson, D. (2021) 'Facebook's Week of Shame', *The Sunday Times*, 19 September, p 31.

Foucault, M. (1979 [1976]) *The History of Sexuality Volume 1: An Introduction*, London: Allen Lane.

Fox, J. and Potocki, B. (2016) 'Lifetime Video Game Consumption, Interpersonal Aggression, Hostile Sexism, and Rape Myth Acceptance: A Cultivation Perspective', *Journal of Interpersonal Violence*, 31(10): 1912–1931.

Fox, S. (2013) '51% of U.S. Adults Bank Online', *Pew Research Centre*, [online] 7 August. Available from: www.pewresearch.org/internet/2013/08/07/51-of-u-s-adults-bank-online/ [Accessed 30 November 2021].

Freedman, D. (2012) 'The Phone Hacking Scandal: Implications for Regulation', *Television and New Media*, 13(1): 17–20.

Friis, S. (2015) 'Beyond Anything We Have Ever Seen: Beheading Videos and the Visibility of Violence in the War Against ISIS', *International Affairs*, 91(4): 725–746.

Fuchs, C. (2011) 'New Media, Web 2.0 and Surveillance', *Sociology Compass*, 5(2): 134–147.

Fuchs, C. (2021) *Social Media: A Critical Introduction*, London: Sage.

Fuller, S. (1997) *Science*, Buckingham: Open University Press.

Furedi, F. (1997) *Culture of Fear: Risk-taking and the Morality of Low Expectation*, London: Continuum.

Furedi, F. (2007) *Invitation to Terror: The Expanding Empire of the Unknown*, London: Continuum.

Gabbiadini, A., Riva P., Andrighetto, L., Volpato C. and Bushman, B.J. (2016) 'Acting like a Tough Guy: Violent-Sexist Video Games, Identification with Game Characters, Masculine Beliefs, and Empathy for Female Violence Victims', *PLOS ONE*, 11(4): 1–5.

Galloway, A. (2004) *Protocol: How Control Exists After Decentralization*, Cambridge, MA: MIT Press.

Galloway, A. (2005) 'Global Networks and the Effects on Culture', *The Annals of the American Academy*, 597(January): 19–31.

Gangadharam, S. (2015) 'The Downside of Digital Inclusion: Expectations and Experiences of Privacy and Surveillance Among Marginal Internet Users', *New Media and Society*, 19(4): 597–615.

Garton Ash, T. (2016) *Free Speech: Ten Principles for a Connected World*, London: Atlantic Books.

Gates, B. (1976) 'An Open Letter to Hobbiests', *Genius*, [online]. Available from: https://genius.com/Bill-gates-an-open-letter-to-hobbyists-annotated [Accessed 12 December 2021].

Gerfert, A. (2018) 'Fake News: A Definition', *Informal Logic*, 28(1): 84–117.

Gerry, F., Muraszkiewicz, J. and Vavoula, N. (2016) 'The Role of Technology in the Fight Against Human Trafficking: Reflections on Privacy and Data Protection Concerns', *Computer Law and Security Review*, 32(2): 205–217.

Giddens, A. (1992) *The Transformation of Intimacy*, Cambridge: Polity Press.

Gilbert, J. (2012) 'Capitalism, Creativity and the Crisis in the Music Industry', opendemocracy.net, [online] 14 September. Available from: www.open democracy.net/en/opendemocracyuk/capitalism-creativity-and-crisis-in-music-industry/ [Accessed 27 March 2022].

Gilmour, S. (2014) 'Policing Crime and Terrorism in Cyberspace: An Overview', *European Review of Organised Crime*, 1(1): 143–159.

Glasgow Media Group (1976) *Bad News*, London: Routledge and Kegan Paul.

Goffman, E. (1952) *The Presentation of Self in Everyday Life*, New York: Doubleday.

Goode, E. and Ben Yahuda, N. (2009) *Moral Panics: The Social Construction of Deviance* (3rd edn), Chichester: Wiley-Blackwell.

Goode, L. (2009) 'Social News, Citizen Journalism and Democracy', *New Media and Society*, 11(8): 1287–1305.

Goode, L. (2015) 'Anonymous and the Political Ethics of Hacktivism', *Popular Communication*, 13(1): 74–86.

Grabosky, P., Smith, R.G. and Dempsey, G. (2001) *Electronic Theft: Unlawful Acquisition in Cyberspace*, Cambridge: Cambridge University Press.

Green, S. (2012) *13 Ways to Steal a Bicycle*, Cambridge, MA: Harvard University Press.

Greenberg, J. (2015) 'Why Facebook and Twitter Can't Just Wipe Out ISIS Online', *Wired Magazine*, [online] 21 November. Available from: www.wired.com/2015/11/facebook-and-twitter-face-tough-choices-as-isis-exploits-social-media/ [Accessed 28 March 2022].

Greenwald, G. (2014) *No Place to Hide: Edward Snowden, the NSA and the Surveillance State,* London: Penguin.

Groth, N. and Birnbaum, J. (1979) *Men who Rape: The Psychology of the Offender,* New York: Plenum Press.

Habermas, J. (1992[1962]) *The Structural Transformation of the Public Sphere,* Cambridge: Polity Press.

Hadjimatheou, K. (2016) 'Surveillance Technologies, Wrongful Criminalisation, and the Presumption of Innocence', *Philosophy and Technology,* 30(1): 39–54.

Haggerty, K. and Ericson, R. (2000) 'The Surveillant Assemblage', *British Journal of Sociology,* 51(4): 605–622.

Halder, D. and Jaishanker, K. (2013) 'Revenge Porn by Teens in the United States and India: A Socio-Legal Analysis', *International Annals of Criminology,* 51(1–2): 85–111.

Hall, S., Critcher, C., Jefferson, T., Clarke, J. and Roberts, B. (1978) *Policing the Crisis: Mugging, the State and Law and Order,* London: Macmillan.

Hampson, N.C.N. (2012) 'Hacktivism: A New Breed of Protest in a Networked World', *Boston College International and Comparative Law Review,* 35(2): 511–542.

Harcup, T. and O'Neill, D. (2017) 'What Is News? News Values Revisited (Again)', *Journalistic Studies,* 18(12): 1470–1488.

Hart, C.B. (2017) 'Free to Play?', in C.B. Hart (ed) *The Evolution and Social Impact of Video Game Economics,* London: Lexington Books, pp 61–80.

Hempill, S.A., Tollit, M. and Kotevski, A. (2014) 'Predictors of Traditional and Cyber-Bullying Victimization', *Journal of Interpersonal Violence,* 30(15): 1–24.

Herman, E. and Chomsky, N. (1995) *Manufacturing Consent: The Political Economy of the Mass Media,* London: Vintage.

Herring, S.C. (2003) 'Gender and power in online communication', in J. Holmes and M. Meyerhoff (eds) *The Handbook of Language and Gender,* Oxford: Blackwell Publishers, pp 202–228.

Hershel, R. and Miori, V.M. (2017) 'Ethics and Big Data', *Technology and Society,* 49(6): 31–36.

Herzog, S. (2011) 'Revisiting the Estonian Cyber-attacks: Digital Threats and Multinational Responses', *Journal of Strategic Security,* 4(2): 49–60.

Hesmondhalgh, D. and Meier, L.M. (2018) 'What the Digitalisation of Music Tells Us About Capitalism, Culture and the Power of the Information Technology Sector', *Information, Communication and Society,* 21(11): 1555–1570.

Hess, C. and Ostrom, E. (2011) *Understanding Knowledge as a Commons,* Cambridge: Massachusetts Institute of Technology Press.

Himanen, P. (2001) *The Hacker Ethic and the Spirit of the Information Age,* London: Secker and Warburg.

Hinduja, S. (2012) 'General Strain, Self-Control and Music Piracy', *International Journal of Cyber Criminology*, 6(1): 951–967.

HM Government (2016) National Cyber Security Strategy, 2016–2021, *GOV.UK*, [online] 11 November. Available from: www.gov.uk/government/publications/national-cyber-security-strategy-2016-to-2021 [Accessed 28 March 2022].

Hogan, B. (2010) 'The Presentation of Self in the Age of Social Media', *Bulletin of Science, Technology and Society*, 30(6): 377–386.

Holbert, R. (2005) 'A Typology for the Study of Entertainment Television Politics', *The American Behavioural Scientist*, 49: 436–453.

Holmes, B. (2003) 'The Emperor's Sword: Art Under WIPO', *World-Information.Org*, [online] 10 December. Available from: http://future-nonstop.org/c/681c81c7823bdfbef214833ac6e0ed30 [Accessed 9 December 2021].

Holt, T. (2012) 'Exploring the Intersections of Technology, Crime and Terror', *Terrorism and Political Violence*, 24(2): 337–354.

Holt, T., Bossler, A.M. and Seigfried-Spellar, K.C. (2017) *Cybercrime and Digital Forensics: An Introduction*, New York: Routledge.

Hornle, J. (2011) 'Countering the Dangers of Online Pornography – Shrewd Regulation of Lewd Content', *European Journal of Law and Technology*, 2(1): 1–26.

Hua, J. and Shaw, R. (2020) 'Corona Virus (COVID-19) "Infodemic"', *International Journal of Environmental Research and Public Health*, 17(7): 1–12.

Igo, S. (2018) *The Known Citizen: The History of Privacy in Modern America*, Cambridge, MA: Harvard University Press.

Jane, E.A. (2016) 'Online Misogyny and Feminist Digilantism', *Continuum*, 30(3): 284–297.

Jansen, S. and Martin, B. (2015) 'The Streisand Effect and Censorship Backfire', *International Journal of Communication*, 9: 656–671.

Jarvis, L. and MacDonald, S. (2015) 'What is Cyberterrorism? Findings From a Survey of Researchers', *Terrorism and Political Violence*, 27(4): 657–678.

Jarvis, L., McDonald, S. and Nouri, L. (2014) 'The Cyberterrorism Threat: Findings from a Survey of Researchers', *Studies in Conflict and Terrorism*, 37(1): 68–90.

Jarvis, L., MacDonald, S. and Chen, T.M. (eds) (2015) *Terrorism Online: Politics, Law and Technology*, London: Routledge.

Jarvis, L., MacDonald, S. and Whiting, A. (2016a) 'Unpacking Cyberterrorism Discourse', *European Journal of Internet Security*, 2(1): 64–87.

Jarvis, L., MacDonald, S. and Whiting, A. (2016b) 'Analogy and Authority in Cyberterrorism Discourse: An Analysis of Global News Coverage', *Global Society*, 30(4): 605–623.

Jewkes, Y. and Andrews, C. (2005) 'Policing the Filth: The Problems of Investigating Online Child Pornography in England and Wales', *Policing and Society*, 15(1): 42–62.

Johns, A. (2009) *Piracy: The Intellectual Property Wars from Gutenberg to Gates*, Chicago, IL: University of Chicago Press.

Johnson, K. (2018) 'Where Did You Get That Story?: An Examination of Story Sourcing Practices and Objectivity on Citizen Journalism Websites', *Electronic News*, 12(3): 165–178.

Jones, C. (2019) 'Fearing for His Life', Longform, [online] 5 May. Available from: https://longform.org/posts/fearing-for-his-life [Accessed 28 March 2022].

Jordan, T. (2008) *Hacking*, Cambridge: Polity Press.

Jordan, T. (2017) 'A Genealogy of Hacking', *Convergence*, 23(5): 528–544.

Jordan, T. and Taylor, P. (2004) *Hacktivism and Cyberwars: Rebels with a Cause?*, London: Routledge.

Kamenetz, A. (2013) 'Why Video Games Succeed Where The Movie and Music Industries Fail', *Fast Company*, [online] 11 July. Available from: www.fastcompany.com/3021008/why-video-games-succeed-where-the-movie-and-music-industries-fail [Accessed 12 December 2021].

Kantor, J. and Twohey, M. (2019) *She Said*, London: Bloomsbury.

Katz, E. and Lazersfeld, P. (1955) *Personal Influence: The Part Played by People in the Flow of Mass Communication*, New York: The Free Press.

Keene, S. (2011) 'Terrorism and the Internet: A Double Edged Sword', *Journal of Money Laundering Control*, 14(4): 359–370.

Kenney, M. (2015) 'Cyber-terrorism in a Post-Stuxnet World', *Orbis*, 59(1): 111–128.

Kenny, K. and Fotaki, M. 'The Costs and Labour of Whistleblowing: Bodily Vulnerability and Post-disclosure Survival', *Journal of Business Ethics*, 182: 341–364.

Khadarova, I. and Pantti, M. (2016) 'Fake News: The Narrative Battle Over the Ukraine Conflict', *Journalistic Practice*, 10(7): 891–901.

Khamis, S. (2017) 'Self-branding, "Micro-celebrity" and the Rise of Social Media Influencers', *Celebrity Studies*, 8(2): 191–208.

Kirkpatrick, G. (2002) 'The Hacker Ethic and the Spirit of the Information Age', *Max Weber Studies*, 2(2): 163–185.

Kirton, A. and David, M. (2013) 'The Challenge of Unauthorized Online Streaming to the English Premier League and Television Broadcasters', in B. Hutchins and D. Rowe (eds) *Digital Media Sport: Technology, Power and Culture in the Network Society*, New York: Routledge, pp 81–94.

Kittler, F. (1997) *Literature, Media, Information Systems*, London: Routledge.

Klein, A.G. (2015) 'Vigilante Media: Unveiling Anonymous and the Hacktivist Persona in the Global Press', *Communication Monographs*, 82(3): 379–401.

Kowalski, R. and Lumber, S. (2013) 'Psychological, Physical and Academic Correlates of Cyberbullying and Traditional Bullying', *Journal of Adolescent Health*, 53(1 supplement): 13–20.

Kowalski, R., Lumber, S. and Agatston, P. (2012) *Cyberbullying: Bullying in the Digital Age*, Oxford: Blackwell.

Krajewski, J.M. and Ekdale, B. (2017) 'Constructing Cholera: CNN iReport, the Haitian cholera epidemic and the limits of citizen journalism', *Journalistic Practice*, 11(2–3): 229–246.

Krone, T. (2004) 'A Typology of Online Child Pornography Offending', *Trends and Issues in Crime and Criminal Justice*, No. 279, Canberra: Australian Institute of Criminology.

Krone, T. (2005) 'International Police Operations Against Online Child Pornography', *Crime and Justice International*, 21(89):11–20.

Krueger, A. and Connolly, M. (2006) 'Rockonomics: The Economics of Popular Music', in V. Ginsberg and D. Throsby (eds) *Handbook of the Economics of Art and Culture*, Amsterdam: North-Holland, pp 667–720.

Kühn, S., Kugler, D., Schmalen, K., Weichenberger, M., Witt, C. and Gallinat, J. (2018) 'The Myth of Blunted Gamers: No Evidence for Desensitization in Empathy for Pain after a Violent Video Game Intervention in a Longitudinal fMRI Study on Non-Gamers', *Neurosignals*, 26(1): 22–30.

Kühn, S., Kugler, D.T., Schmalen, K., Weichenberger, M., Witt, C. and Gallinat, J. (2019) 'Does Playing Violent Video Games Cause Aggression? A Longitudinal Intervention Study', *Molecular Psychiatry*, 24(8): 1220–1234.

Lachow, I. (2009) 'Cyber Terrorism: Menace or Myth', in *Cyberpower and National Security*, Washington, DC: National Defence University Press, pp 434–467.

LaCroix, J.M., Burrows, C.N. and Blanton, H. (2018) 'Effects of Immersive, Sexually Objectifying, and Violent Video Games on Hostile Sexism in Males', *Communication Research Reports*, 35(5): 413–423.

Lanning, K. (2018) 'The Evolution of Grooming: Concept and Form', *Journal of Interpersonal Violence*, 30(1): 5–16.

Lastowka, G. (2015) 'Copyright Law and Video Games: A Brief History of an Interactive Medium', in M. David and D. Halbert (eds) *The Sage Handbook of Intellectual Property*, London: Sage. pp 495–514.

Latour, B. (2005) *Reassembling the Social: An Introduction to Actor-Network Theory*, Oxford: Oxford University Press.

Lavorgna, A. (2020) *Cybercrimes: Critical Issues*, London: Macmillan.

Lee, J. (2015) 'Non-profits in the Commons Economy', in M. David and D. Halbert (eds) *The Sage Handbook of Intellectual Property*, London: Sage, pp 335–354.

Leopold, L. and Bell, M.P. (2017) 'News Media and the Socialisation of Protest: An Analysis of Black Lives Matter Articles', *Equality, Diversity and Inclusion*, 36(8): 720–735.

Lessig, L. (2002) *The Future of Ideas*, New York: Vintage.

Lessig, L. (2004) *Free Culture*, New York: Penguin Press.

Levi, M., Doig, A., Gundur, R., Wall, D. and Williams, M. (2017) 'Cyberfraud and the Implications for Effective Risk-based Responses: Themes from UK Research', *Crime, Law and Social Change*, 67: 77–96.

Lewis, R., Sharp, E., Remnant, J. and Redpath, R. (2015) '"Safe Spaces": Experiences of Feminist Women-only Space', *Sociological Research Online*, 20(4): 9.

Lewis, R., Rowe, M. and Wiper, C. (2016) 'Online Abuse of Feminists as an Emerging Form of Violence Against Women and Girls', *British Journal of Criminology*, 57(6): 1462–1481.

Liebler, R. (2015) 'Copyright and Ownership of Fan Created Works: Fanfiction and Beyond', in M. David and D. Halbert (eds) *The Sage Handbook of Intellectual Property*, Sage, London, pp 391–403.

Lindgren, S. and Lundstrom, R. (2011) 'Pirate Culture and Hacktivism Mobilization: The Cultural and Social Protocols of Wikileaks on Twitter', *New Media and Society*, 13(6): 999–1018.

Lindner, A. (2017) 'Editorial Gatekeeping in Citizen Journalism', *New Media and Society*, 19(8): 1177–1193.

Litman, J. (1991), 'Copyright as Myth', *Pittsburgh Law Review*, 53(1): 235–249.

Logan, T. and Walker, R. (2017) 'Stalking: A Multidimensional Framework for Assessment and Safety Planning', *Trauma, Violence and Abuse*, 18(2): 200–222.

Longshaw, M. (2011) 'Software Piracy: The Greatest Threat to the Gaming Industry?', Digital Spy, [online] 27 November. Available from: www.digital spy.co.uk/gaming/news/a352906/software-piracy-the-greatest-threat-to-the-gaming-industry.html [Accessed 12 December 2021].

Love, C. (2000) 'Courtney Love Does the Math', *Salon*, [online] 14 June. Available from: www.salon.com/2000/06/14/love_7/ [Accessed 9 December 2021].

Luhmann, N. (1996) *Social Systems*, Redwood City, CA: Stanford University Press.

Lundborg, T. (2016) 'The Virtualisation of Security: Philosophies of Capture and Resistance in Baudrillard, Agamben and Deleuze', *Security Dialogue*, 47(3): 255–270.

Lynch, K.P. and Logan, T. (2015) 'Police Officers' Attitudes and Challenges with Charging Stalking', *Violence and Victims*, 30(6): 1037–1048.

Lyon, D. (2014) 'Surveillance, Snowden and Big Data: Capacities, Consequences, Critique', *Big Data and Society*, 1(2): 1–13.

Ma, L., Montgomery, A., Singh, P.V. and Smith, M.D. (2014) 'An Empirical Analysis of the Impacts of Pre-release Movie Piracy on Box Office Revenue', *Information Systems Research*, 25(3): 590–603.

Mader, C. (2012) *The Passion of Bradley Manning*, London: Verso.

Malinowski, B. (2002 [1922]) *Argonauts of the Western Pacific*, Abingdon: Routledge.

Man, J. (2002) *Gutenburg: One Man Who Remade the World with Words*, New York: Wiley.

Maple, C. (2017) 'Security and Privacy in the Internet of Things', *Journal of Cyber Policing*, 2(2): 155–184.

Marchi, R. (2012) 'With Facebook, Blogs and Fake News, Teens Reject Journalistic "objectivity"', *Journal of Communications Inquiry*, 36(3): 386–392.

Marcum, C.D., Ricketts, M. and Higgins, G. (2010) 'Assessing Sex Experiences of Online Victimisation: An Examination of Adolescent Online Behaviour Using Routine Activity Theory', *Criminal Justice Review*, 35(4): 412–437.

Marcuse, H. (1992[1955]) *Eros and Civilisation: A Philosophical Inquiry into Freud*, Boston, MA: Beacon Press.

Marganski, A. and Melander, L. (2015) 'Intimate Partner Violence, Victimization in the Cyber and Real World', *Journal of Interpersonal Violence*, 33(7): 1071–1095.

Marshall, L. (2005) *Bootlegging*, London: Sage.

Marshall, L. (2013) 'The 360 Deal and the "New" Music Industry', *European Journal of Cultural Studies*, 16(1): 77–99.

Marshall, L. (2015) '"Let's Keep Music Special, F-Spotify": On-demand Streaming and the Controversy Over Artist Royalties', *Creative Industry Journal*, 8(2), 177–189.

Mason, C. and Magnet, S. (2012) 'Surveillance Studies and Violence Against Women', *Surveillance and Society*, 10(2), 105–118.

Matza, D. (1964) *Delinquency and Drift*, New York: Wiley.

Maurushat, A. (2013) 'From Cybercrime to Cyberwar: Security Through Obscurity or Security Through Absurdity?', *Canadian Foreign Policy Journal*, 19(2): 119–122.

Mauss, M. (1990 [1925]) *The Gift*, Abingdon: Routledge.

Maxwell, L. (2018) 'Whistleblower, Traitor, Soldier, Queer? The Truth of Chelsea Manning', *Yale Review*, 10(1): 97–107.

May, C. and Sell, S. (2005) *Intellectual Property Rights*, Boulder, CO: Lynne Rienner.

McCormack, M. (2012) *The Declining Significance of Homophobia: How Teenage Boys are Redefining Masculinity and Heterosexuality*, Oxford: Oxford University Press.

McCormack, M. and Wignall, L. (2017) 'Enjoyment, Exploration and Education: Understanding the Consumption of Pornography among Young Men with Non-Exclusive Sexual Orientations', *Sociology*, 51(5): 975–991.

McDermott, L. (1996) 'Self-Representation in Upper Paleolithic Female Figurines', *Current Anthropology*, 37(2): 227–275.

McDermott, Y. (2017) 'Conceptualizing the Right to Data Protection in an Era of Big Data', *Big Data and Society*, 4(1): 1–7.

McGlynn, C. and Rackley, E. (2009) 'Criminalising Extreme Pornography: A Lost Opportunity', *The Criminal Law Review*, 4: 245–260.

McGlynn, C., Rackley, E., Johnson, K., Henry, N., Flynn, A., Powell, A. and Gavey, N. and Scott, A. (2019) *Shattering Lives and Myths: A Report on Image-based Sexual Abuse*, Project Report. Durham University; University of Kent, [online]. Available from: https://claremcglynn.files. wordpress.com/2019/06/shattering-lives-and-myths-final.pdf [Accessed 26 February 2023].

McGonagle, T. (2010) 'Minorities and Online Hate Speech: A Parsing of Selected Complexities', *European Yearbook of Minority Issues*, 9: 419–440.

McKee, A. (2010) 'Does Pornography Harm Young People?', *Australian Journal of Communications*, 37(1): 17–36.

Merton, R.K. (1942/1972) 'The Institutional Imperatives of Science', in B. Barnes (ed) *Sociology of Science*, London: Penguin, pp 65–79.

Mill, J.S. (2008[1859]) *On Liberty*, Oxford: Oxford University Press.

Miller, L. (2012) 'Stalking: Patterns, Motives, and Intervention Strategies', *Aggression, and Violent Behaviour*, 17(6): 495–506.

Millward, P. (2017) *A Whole New Ball Game: The English Premier League and Television Broadcast Rights*, London: Routledge.

Min, S. (2016) 'Conversation Through Journalism: Searching for Organising Principles of Public and Citizen Journalism', *Journalism*, 17(5): 567–582.

Mishna, F., Khoury-Kassabri, M., Gadalla, T. and Daciuk, J. (2012) 'Risk Factors for Involvement in Cyber Bullying: Victims, Bullies and Bully-victims', *Children and Youth Services Review*, 34(1): 63–70.

Moore, A.D. (2011) 'Privacy, Security, and Government Surveillance: WikiLeaks and the New Accountability', *Public Affairs Quarterly*, 25(2): 141–156.

Moore, D. and Rid, T. (2016) 'Cryptopolitik and the Darknet', *Survival*, 58(1): 7–38.

Moore, R. (2015) *Cybercrime: Investigating High-Technology Computer Crime*, New York: Routledge.

Morris, R. and Higgens, G. (2010) 'Criminological Theory in the Digital Age: The Case of Social Learning Theory and Digital Piracy', *Journal of Criminal Justice*, 38(4): 470–480.

Mould, T. (2018) 'A Doubt-centred Approach to Contemporary Legend and Fake-news', *The Journal of American Folklore*, 131(522): 413–420.

Moyo, L. (2011) 'Blogging Down a Dictatorship: Human Rights, Citizen Journalists and the Right to Communicate in Zimbabwe', *Journalism*, 12(6): 745–760.

Mrah, I. (2019) 'Citizen Journalism in the Digital Age: The Case of the 2011 Social Protests in Egypt', *Journal of Sociology and Anthropology*, 3(1): 1–10.

Nagle, A. (2017) *Kill all Normies*, London: Zero Books.

Nair, A. (2010) 'Real Porn and Pseudo Porn: The Regulatory Road', *International Review of Law, Computers and Technology*, 34(3): 223–232.

Navaro, J.N. and Jasinksi, J.L. (2012) 'Going Cyber: Using RAT to Predict Cyberbullying Experiences', *Sociological Spectrum*, 32(1): 81–94.

Neff, G. and Nafus, D. (2016) *Self-Tracking*, Cambridge, MA: MIT Press.

Neiborg, D. (2016) 'From Premium to Freemium: The Political Economy of the App', in M. Williams and T. Leaver (eds) *Social, Casual and Mobile Games: The Changing Gaming Landscape*, New York: Bloomsbury, pp 225–240.

Nelson, J. and Teneja, H. (2018) 'The Small, Disloyal Fake News Audience', *New Media and Society*, 20(10): 3720–3737.

Ngo, F.T. (2018) '"Stalking": An Examination of the Correlates of Subsequent Police Responses', *Policing*, 42(3): 362–375.

Nguyen, A. and Scifo, S. (2018) 'Mapping the Citizen News Landscape', in T.P. Vos (ed) *Journalism*, Berlin: De Gruyter Mouton, pp 373–390.

Nikitina, S. (2012) 'Hackers as Tricksters of the Digital Age: Creativity in Hacker Culture', *The Journal of Popular Culture*, 45(1): 133–152.

Nissenbaum, H. (2004) 'Hackers and the Contested Ontology of Cyberspace', *New Media and Society*, 6(2): 195–217.

Nissenbaum, H. (2010) *Privacy in Context: Technology, Policy and the Integrity of Social Life*, Stanford, CA: Stanford University Press.

Nissenbaum, H. (2011) 'A Contextual Approach to Privacy Online', *Daedalus*, 140(4): 32–48.

Nobles, M., Reyns, B., Fox, K. and Fisher, B. (2012) 'Protection Against Pursuit: A Conceptual and Empirical Comparison of Cyberstalking and Stalking Victimization Among a National Sample', *Justice Quarterly*, 31(6): 986–1014.

Nycyk, M. (2015) 'Adult-to-adult Cyberbullying: An Exploration of a Dark Side of the Internet', Publisher: Michael Nycyk, Brisbane, [online]. Available from: www.academia.edu/11836687/Adult-to-Adult_ Cyberbullying_An_Exploration_of_a_Dark_Side_of_the_Internet [Accessed 28 March 2022].

Ohlin, J., Govern, K. and Finkelstein, C. (2015) *Cyberwar*, Oxford: Oxford University Press.

Olesen, T. (2018) 'The Democratic Drama of Whistle Blowing', *European Journal of Social Theory*, 21(4): 508–525.

Olesen, T. (2019) 'The Politics of Whistleblowing in Digital Societies', *Politics and Society*, 47(2): 277–297.

ONS (Office for National Statistics) (2016) Overview of Fraud Statistics: Year Ending March 2016, ONS, [online] 21 July. Available from: www.ons.gov. uk/peoplepopulationandcommunity/crimeandjustice/articles/overview offraudstatistics/yearendingmarch2016 [Accessed 28 November 2021].

Ost, S. (2010) 'Criminalising Fabricated Images of Child Pornography: A Matter of Harm or Morality?', *Legal Studies*, 30(2): 230–256.

Owens, J.G. (2016) 'Why Definitions Matter: Stalking Victimization in the US', *Journal of Interpersonal Violence*, 31(12): 2196–2226.

Pabian, S. and Vandebosch, H. (2016) 'An Investigation of Short-term Longitudinal Associations Between Social Anxiety and Victimization and Perpetration of Traditional Bullying and Cyberbullying', *Journal of Youth and Adolescence*, 45(2): 328–339.

Pariser, E. (2014[2011]) 'Beware Online "Filter Bubbles"', *Ted Talk*, [online] 1 February. Available from: https://tedsummaries.com/2014/02/01/eli-pariser-beware-online-filter-bubbles/ [Accessed 23 November 2021].

Park, Y.J. (2013) 'Digital Literacy and Privacy Behaviour Online', *Communication Research*, 40(2): 215–236.

Park, Y.J. (2015) 'Do Men and Women Differ in Privacy? Gendered Privacy and (In)equality in the Internet', *Computers and Human Behaviour*, 50: 252–258.

Perry, B. and Olsson, P. (2009) 'Cyberhate: The Globalization of Hate', *Information and Communication Technology and Law*, 18(2): 185–199.

Petella-Rey, P. (2018) 'Beyond Privacy: Bodily Integrity as an Alternative Framework for Understanding Non-consensual Pornography', *Information, Communication and Society*, 21(5): 786–791.

Phipps, A. and Young, I. (2015) '"Lad Culture" in Higher Education', *Sexualities*, 18: 459–479.

Pidduck, J. (2010) 'Citizen Journalism in Burma and the Legacy of Graham Spry', *Canadian Journal of Communication*, 35(3): 473–485.

Pittaro, M. (2007) 'Cyber Stalking: An Analysis of Online Harassment and Intimidation', *International Journal of Cyber Criminology*, 1(2): 180–197.

Pomerantsev, P. (2016) *Nothing is True and Everything is Possible*, London: Faber and Faber.

Poster, M. (2006) *Information Please: Culture and Politics in the Age of Digital Media*, Durham, NC: Duke University Press.

Potter, R.H. and Potter, L.A. (2001) 'The Internet, Cyberporn, and Sexual Exploitation of Children: Media Moral Panics and Urban Myths for Middle-class Parents?', *Sexuality and Culture*, 5(3): 31–48.

Potts, L. (2012) 'Amanda Palmer and the #LOFNOTC: How Online Fan Participation is Rewriting Music Labels', *Participations: Journal of Audience and Reception Studies*, 9(2): 360–382.

Powell, K.J. (2016) 'Making #Black Lives Matter: Michael Brown, Eric Garner, and the Specters of Black Life – Toward a Hauntology of Blackness', *Cultural Studies ↔ Critical Methodologies*, 16(3): 253–260.

Pratt, T.C., Holtfreter, K. and Reisig, M.D. (2010) 'Routine Online Activity and Internet Fraud Targeting: Extending the Generality of Routine Activity Theory', *Journal of Research in Crime and Delinquency*, 47(3): 267–296.

Presdee, M. (2000) *Cultural Criminology and the Carnival of Crime*, London: Routledge.

Qin, J. (2015) 'Hero on Twitter, Traitor on News: How Social Media and Legacy News Frame Snowden', *International Journal of Press/Politics*, 20(2): 166–184.

Rege, A. (2009) 'What's Love Got to Do with It? Exploring Online Dating Scams and Identity Fraud', *International Journal of Cyber Criminology*, 3(2): 494–512.

Reijnen, L., Bulten, E. and Nijman, H. (2009) 'Demographic and Personality Characteristics of Internet Child Pornography Downloaders in Comparison to Other Offenders', *Journal of Child Sexual Abuse*, 18(6): 611–622.

Ressa, M. (2022) *How to Stand Up to a Dictator*, London: Virgin Books.

Reyns, B.W. (2013) 'Online Routines and Identity Theft Victimization: Further Expanding Routine Activity Theory Beyond Direct-contact Offences', *Journal of Research in Crime and Delinquency*, 50(20): 216–238.

Reyns, B.W., Henson, B. and Fisher, B.S. (2011) 'Being Pursued Online: Applying Cyberlifestyle-Routine Activities Theory to Cyberstalking Victimization', *Criminal Justice and Behavior*, 38(11): 1149–1169.

Ribbens, W. and Malliet, S. (2015) 'How Male Young Adults Construe Their Playing Style in Violent Video Games', *New Media & Society*, 17(10): 1624–1642.

Rice, E., Petering, R., Rhoades, H., Winetrobe, H., Goldbach, J. Plant, A., Montoya, J. and Kordic, T. (2015) 'Cyberbullying Perpetration and Victimization Among Middle-school Students', *American Journal of Public Health*, 105(3): 66–72.

Rifkin, J. (2014) *The Zero Marginal Cost Society*, New York: Palgrave Macmillan.

Roberts, A. and Dziegielewski, S. (1996) 'Assessment Typology and Intervention with the Survivors of Stalking', *Aggression and Violent Behaviour*, 1(4): 359–368.

Rogers, J. (2013) *The Death and Life of the Music Industry in the Digital Age*, London: Bloomsbury.

Rojek, C. (2016) *Presumed Intimacy*, Cambridge: Polity Press.

Rosen, J. (2013) 'The Delete Squad: Google, Twitter, Facebook and the New Global Battle Over the Future of Free Speech', *New Republic*, [online] 29 April. Available from: https://newrepublic.com/article/113045/free-speech-internet-silicon-valley-making-rules [Accessed 28 March 2022].

Roser, M. and Ritchie, H. (2013) 'Homicides', *OurWorldInData.org* [online] July, Available from: https://ourworldindata.org/homicides [Accessed 1 June 2022].

Rotte, D. and Steinmetz, K. (2013) 'The Case of Bradley Manning', *Contemporary Justice Review*, 16(2): 280–293.

Royle, A. (2013) 'Pirates Ahoy! Copyright and Internet File-Sharing,' *North East Law Review*, 1(4): 51–79.

Rule, T. (2019) 'Contextual Integrity and its Discontents: A Critique of Helen Nissenbaum's Normative Arguments', *Policy and Internet*, 11(3): 60–79.

Salami, I. (2017) 'Terrorism financing with virtual currencies: Can Regulatory Technology Solutions Combat This?', *Studies in Conflict and Terrorism*, 41(12): 968–989.

Sandell, R. (2007) 'Off the Record', *Prospect Magazine*, [online] 137, 1 August. Available from: www.prospectmagazine.co.uk/magazine/offthe record [Accessed 27 March 2022].

Scheuerman, W. (2014) 'Whistleblowing as Civil Disobedience: The Case of Edward Snowden', *Philosophy and Social Criticism*, 40(7): 609–628.

Schwartz, J.A. (2014) *Online File Sharing: Innovations in Media Consumption*, Abingdon: Routledge.

Shoemaker, P. and Reese, S. (2014) *Mediating the Message in the 21st Century*, New York: Routledge.

Short, E., Stanley, T., Baldwin, M. and Scott, G. (2015) 'Behaving Badly Online: Establishing Norms of Unacceptable Behaviour', *Studies in Media and Communication*, 3(1): 1–10.

Silbey, J. (2015) *The Eureka Myth: Creators, Innovators, and Everyday Intellectual Property*, Stanford, CA: Stanford University Press.

Simmel, G. (1906) 'The Sociology of Secrecy and of the Secret Societies', *American Journal of Sociology*, 11(4): 441–498.

Simons, G. (2014) 'The Impact of Social Media and Citizen Journalism on Mainstream Russian News', *Russian Journal of Communication*, 8(1): 33–51.

Singer, P.W. and Friedman, A. (2014) *Cybersecurity and Cyberwar: What Everyone Needs to Know*, Oxford: Oxford University Press.

Smith, S.G., Chen, J., Basile, K.C., Gilbert, L.K., Merrick, M.T., Patel, N., Walling, M. and Jain, A. (2017) *The National Intimate Partner and Sexual Violence Survey (NISVS): 2010–2012 State Report*, Atlanta: National Center for Injury Prevention and Control, Centers for Disease Control and Prevention.

Snowden, E. (2019) *Permanent Record*, London: Macmillan.

Söderberg, J. (2008) *Hacking Capitalism: The Free and Open Source Software Movement*, New York: Routledge.

Söderberg, J. (2013) 'Determining Social Change: The Role of Technological Determinism in the Collective Action Framing of Hackers', *New Media and Society*, 15(8): 1277–1293.

Southworth, C., Finn, J., Dawson, S., Fraser, C. and Tucker, S. (2007) 'Intimate Partner Violence, Technology and Stalking', *Violence Against Women*, 13(8): 842–856.

Spokes, M. (2018) '"War … War Never Changes": Exploring Explicit and Implicit Encounters with Death in a Post-Apocalyptic Gameworld', *Mortality*, 23(2): 135–150.

Statista (2020) 'Video Game Industry – Statistics and Facts', *Statista*, [online] 19 November. Available from: www.statista.com/topics/868/video-games/ [Accessed 12 December 2021].

Statista (2021) 'Online Banking Penetration in Great Britain from 2007 to 2020', *Statista*, [online] 11 January. Available from: www.statista.com/statistics/286273/internet-banking-penetration-in-great-britain/ [Accessed 30 November 2021].

Statista (2022) 'Software Market Revenue in the World from 2016 to 2021', *Statista*, [online] 11 February. Available from: www.statista.com/forecasts/963597/software-revenue-in-the-world [Accessed 28 March 2022].

Steinmetz, K.F. (2016) *Hacked: A Radical Approach to Hacker Culture and Crime*, New York: New York University Press.

Stermer, S.P. and Burkley, M. (2012) 'Xbox or SeXbox? An Examination of Sexualized Content in Video Games', *Social and Personality Psychology Compass*, 6(7): 525–535.

Stöcker, C. (2011) 'A Dispatch Disaster in Six Acts', *Der Spiegel*, [online] 1 September. Available from: www.spiegel.de/international/world/leak-at-wikileaks-a-dispatch-disaster-in-six-acts-a-783778.html [Accessed 14 December 2022].

Stohl, C. and Stohl, M. (2007) 'Networks of Terror: Theoretical Assumptions and Pragmatic Consequences', *Communication Theory*, 17(2): 93–124.

Stohl, M. (2007) 'Cyber Terrorism: A Clear and Present Danger, the Sum of all Fears, Breaking Point or Patriot Games?', *Crime, Law and Social Change*, 46(4–5): 223–238.

Stranger, A. (2019) *Whistleblowers: Honesty in America from Washington to Trump*, New Haven, CT: Yale University Press.

Strawhun, J., Adams, N. and Huss, T. (2013) 'An Assessment of Cyberstalking, Violence and Victims: An Expanded Examination Including Social Networking, Attachment, Jealousy, and Anger in Relation to Violence and Abuse', *Violence and Victims*, 28(4): 715–730.

Stryker, C. (2012) *Hacking the Future: Privacy, Identity and Anonymity on the Web*, New York: Duckworth.

Sulston, J. and Ferry, G. (2009) *The Common Thread*, London: Corgi Books.

Taliharm, A. (2010) 'Cyberterrorism: in Theory or in Practice?', *Defence Terrorism Review*, 3(2): 59–74.

Tandoc, E.C., Lim, Z.W., Zheng, W.L. and Ling, R. (2017) 'Defining Fake News: A Typology of Scholarly Definitions', *Digital Journalism*, 6(2): 137–153.

Tandoc, E.C., Ling, R., Westlund, O., Duffy, A., Goh, D. and Wei, L. (2018) 'Audiences' Acts of Authentication in the Age of Fake News: A Conceptual Framework', *New Media and Society*, 20(8): 2745–2763.

Tapscott, D. and Williams, A.D. (2008) *Wikinomics*, London: Atlantic Books.

Tavani, H. and Grodzinsky, F. (2014) 'Trust, Betrayal and Whistle-Blowing', *SIGCAS Computers and Society*, 44(3): 8–13.

Taylor, M. (1999) 'The Nature and Dimensions of Child Pornography on the Internet', Paper presented at the *Combating Child Pornography on the Internet conference*, Vienna, Austria, 29 September–1 October 1999.

Tehranian, J. (2016) 'Sanitizing Cyberspace: Obscenity, Miller, and the Future of Public Discourse on the Internet', *Journal of Intellectual Property Law*, 11(1): 1–28.

Tewksbury, D. and Rittenburg, J. (2012) *News on the Internet*, Oxford: Oxford University Press.

Thompson, J.B. (2005) *Books in the Digital Age*, Cambridge: Polity Press.

Thompson, J.B. (2012) *Merchants of Culture*, Cambridge: Polity Press.

Tjaden, P. (1997) 'The Crime of Stalking: How Big Is the Problem?', *National Institute of Justice Research Review*, November: 1–4.

Trottier, D. (2012) 'Policing Social Media', *Canadian Review of Sociology*, 49(4): 411–425.

Turgeman-Goldschmidt, O. (2008) 'Meanings that Hackers Assign to their Being a Hacker', *International Journal of Cyber Criminology*, 2(2): 382–396.

UK Finance (2021) 'Fraud – the Facts 2021: The Definitive Overview of Payment Industry Fraud', *UK Finance*, [online]. Available from: www.ukfinance.org.uk/system/files/Fraud%20The%20Facts%202021-%20FINAL.pdf [Accessed 29 November 2021].

United Nations (2013) 'Comprehensive Study on Cybercrime – Draft Report', *United Nations Office of Drugs and Crime: Vienna*, [online] February. Available from: www.unodc.org/documents/organized-crime/UNODC_CCPCJ_EG.4_2013/CYBERCRIME_STUDY_210213.pdf [Accessed 29 November 2021].

Vaidhyanathan, S. (2003) *Copyrights and Copywrongs: The Rise of Intellectual Property and How It Threatens Creativity*, New York: New York University Press.

Vargo, C.J., Guo, L. and Amazeen, M. (2018) 'The Agenda Setting Power of Fake News', *New Media and Society*, 20(5): 2028–2049.

Varone, A. (2001) *Eroticism in Pompeii*, Los Angeles: Getty Trust Publications: J. Paul Getty Museum.

Vegh, S. (2002) 'Hacktivism or Cyberterrorists? The Changing Media Discourse On Hacking', *First Monday* 7(10), [online] 16 April. Available from: 10.5210/fm.v7i10.998 [Accessed 28 March 2022].

Vera-Gray, F. (2017) '"Talk about a Cunt with Too Much Idle Time": Trolling Feminist Research', *Feminist Review*, 115(1): 61–78.

Vera-Gray, F. (2018) *The Right Amount of Panic: How Women Trade Freedom for Safety*, Bristol: Policy Press.

Vera-Gray, F., McGlynn, C., Kureshi, I. and Butterby, K. (2021) 'Sexual Violence as a Sexual Script in Mainstream Online Pornography', *The British Journal of Criminology*, 61(5): 1243–1260.

Virilio, P. (2000) *The Information Bomb*, London: Verso.

Vitak, J. (2012) 'The Impact of Context Collapse and Privacy on Social Network Site Disclosures', *Journal of Broadcasting and Electronic Media*, 56(4): 451–470.

Vosoughi, S., Roy, D. and Aral, S. (2018) 'The Spread of True and False News Online', *Science*, 359 (6380): 1146–1151.

Waasdorp, T. and Bradshaw, C. (2015) 'The Overlap Between Cyberbullying and Traditional Bullying', *Journal of Adolescent Health*, 56(5): 483–488.

Waldman, A. (2018) *Privacy and Trust: Information Privacy for an Information Age*, Cambridge: Cambridge University Press.

Wall, D.S. (2007) *Cybercrime: The Transformation of Crime in the Information Age*, Cambridge: Polity Press.

Wall, D.S. (2010) 'Micro-Frauds: Virtual Robberies, Stings and Scams in the Information Age', in T. Holt and B. Schell (eds) *Corporate Hacking and Technology-Driven Crime: Social Dynamics and Implications*, Hershey, PA: IGI Global, pp 68–85.

Wall, D. (2012) 'The Devil Drives a Lada: The Social Construction of Hackers as Cybercriminals', in C. Gregoriou (ed) *The Construction of Crime*, Basingstoke: Palgrave, pp 4–18.

Wall, D. (2013) 'Policing Identity Crimes', *Policing and Society*, 23(4): 437–460.

Wall, D. and Williams, M. (2007) 'Policing Diversity in the Global Digital Age: Maintaining Order in Virtual Communities', *Criminology and Criminal Justice*, 7(4): 391–415.

Wall, M. (2015) 'Citizen Journalism', *Digital Journalism* 3(6): 797–813.

Weimann, G. (2005) 'Cyberterrorism: The Sum of All Fears?', *Studies in Conflict and Terrorism*, 28(2): 129–149.

Weimann, G. (2015) *Terrorism in Cyberspace: The Next Generation*, New York: Columbia University Press.

West, S. (2019) 'Data Capitalism', *Business and Society*, 58(8): 20–41.

Whitehouse, G. (2010) 'Newsgathering and Privacy: Expanding Ethics Codes to Reflect Change in the Digital Media Age', *Journal of Mass Media Ethics*, 25(4): 310–327.

Whitty, M. and Buchanan, T. (2016) 'The Online Dating Romance Scam: The Psychological Impact on Victims – Both Financial and Non-financial', *Criminology & Criminal Justice*, 16(2): 176–194.

Wick, S.E., Nagoshi, C., Basham, R., Jordan, C. Kyoung Kim, Y., Phuong Nguyen, A. and Lehmann, P. (2017) 'Patterns of Cyber Harassment and Perpetration among College Students in the United States: A Test of Routine Activity Theory', *International Journal of Cyber Criminology*, 11(1): 24–38.

William, T.J. (2011) 'Intervention: Hacking History, from Analogue to Digital and back', *Rethinking History*, 15(2): 287–296.

Williams, M. (2007) 'Policing and Cybersociety: The Maturation of Regulation within an Online Community', *Policing and Society*, 17(1): 59–82.

Williams, M., Edwards, A., Housley, W. Burnap, P., Rana, O., Avis, N., Morgan, J. and Sloan, L. (2013) 'Policing Cyber-neighbourhoods: Tension Monitoring and Social Media Networks', *Policing and Society*, 23(4): 461–481.

Woodlock, D. (2017) 'The Abuse of Technology in Domestic Violence and Stalking', *Violence Against Women*, 23(5): 584–602.

Wring, D. (2012) '"It's Just Business": The Political Economy of the Hacking Scandal', *Media, Culture and Society*, 34(5): 631–636.

Wykes, M. (2007) 'Constructing Crime: Culture, Stalking, Celebrity and Cyber', *Crime, Media, Culture*, 3(2): 158–174.

Yar, M. (2012) 'Crime, Media and the Will-to-Representation: Reconsidering Relationships in the New Media Age', *Crime, Media, Culture*, 8(3): 245–260.

Yar, M. (2013) 'The Policing of Internet Sex Offenders: Pluralised Governance Versus Hierarchies of Standing', *Policing and Society*, 23(4): 482–497.

Yar, M. and Steinmetz, K. (2019) *Cybercrime and Society* (3rd edn), London: Sage.

Young, J. (1971) *The Drugtakers: The Social Meaning of Drug Use*, London: Paladin.

Zeng, J., Burgess, J. and Bruns, A. (2019) 'Is Citizen Journalism Better Than Professional Journalism in Fact Checking Rumours in China? How Weibo Users Verified Information Following the 2015 Tianjin blasts', *Global Media and China*, 4(1): 13–35.

Zentner, A. (2006) 'Measuring the Effect of File Sharing on Music Purchases', *The Journal of Law and Economics*, 49(1): 63–90.

Zuboff, S. (1988) *In the Age of the Smart Machine*, New York: Basic Books.

Zuboff, S. (2015) 'Big Other: Surveillance Capitalism and the Prospects of an Information Civilization', *Journal of Information Technology*, 30(1): 75–89.

Zuboff, S. (2019) *The Age of Surveillance Capitalism: The Fight for a Human Future at the New Frontier of Power*, New York: Profile Books.

Index

References in **bold** type refer to tables.

Printed and bound by CPI Group (UK) Ltd, Croydon, CR0 4YY

16/04/2025

14658339-0003